INTRODUCTION TO NANOTECHNOLOGY
ASET 201 (TEXTBOOK)

By Boris Zubry

Custom Edition

Community College of Philadelphia

1700 Spring Garden Street,

Philadelphia, Pennsylvania 19130

First Edition

All characters in this book are fictitious and any resemblance to a real person, living or dead, is coincidental.

ISBN-13: 9781089490876

Printed in the United States of America

Zubry Publishing   *- ZP -*   *Princeton, USA*

bzubry@ccp.edu

To students yearning to learn,
To the esteemed colleges eager to teach.

TABLE OF CONTENTS:

INTRODUCTION TO NANOTECHNOLOGY

By Boris Zubry

1.0   Introduction to Nanotechnology

Definition:

Nanotechnology is the management of materials on an atomic or molecular scale expressly to build microscopic and nanoscale devices (such as robots). Placing atoms as though they were bricks, **nanotechnology** will give us complete control over the structure of matter, allowing us to build any substance or structure permitted by the laws of nature.

Nanotechnology is science, engineering, and technology conducted at the nanoscale, which is about 1 to 100 nanometers. Physicist Richard Feynman is considered the father of nanotechnology. Nanoscience and nanotechnology are the study and application of extremely small things and can be used across all the scientific fields, such as chemistry, biology, physics, materials science, healthcare and engineering.

Recommended videos:

https://www.bing.com/videos/search?q=Definition+of+nanotechnology+video&&view=detail&mid=263C623F0635C3C4AA18263C623F0635C3C4AA18&&FORM=VRDGAR

https://www.bing.com/videos/search?q=Definition+of+nanotechnology+video&&view=detail&mid=F07320092D14F2B2E695F07320092D14F2B2E695&&FORM=VRDGAR

Origins of Nanotechnology:

The **history of nanotechnology** goes back the development of the concepts and experimental work falling under the broad category of nanoscience and nanotechnology. Although nanotechnology is a relatively recent development in scientific research, the development of its central concepts happened over a longer period. The emergence of nanotechnology in the 1980s was caused by the merging of experimental advances such as the invention of the **scanning tunneling microscope** in 1981 and the discovery of

fullerenes in 1985, with the elucidation and popularization of a conceptual framework for the goals of nanotechnology beginning with the 1986 publication of the book *Engines of Creation*. The field was subject to growing public awareness and controversy in the early 2000s, with prominent debates about its potential implications as well as the feasibility of the applications envisioned by promoters of molecular nanotechnology. The early 2000s also saw the beginnings of commercial applications of nanotechnology, although these were limited to bulk applications of nanomaterials rather than the transformative applications envisioned by the field.

A scanning tunneling microscope (**STM**) is an instrument for imaging surfaces at the atomic level. Its development in 1981 earned its inventors, Gerd Binnig and Heinrich Rohrer (at IBM Zürich), the Nobel Prize in Physics in 1986.

A **fullerene** is an allotrope of carbon whose molecule consists of carbon atoms connected by single and double bonds forming a closed or partially closed mesh, with fused rings of five to seven atoms. The molecule may be a hollow sphere, ellipsoid, tube, or many other shapes and sizes.

Nanomaterials describe, in principle, materials of which a single unit is sized (in at least one dimension) between 1 to 1000 nanometers (10^{-9} meter) but usually is 1 to 100 nm (the usual definition of nanoscale). For example, a single human hair is about 100,000 nanometers across.

Recommended videos:

https://www.bing.com/videos/search?q=Origins+of+Nanotechnology+video&&view=detail&mid=19A884FE16BF00A51B8A19A884FE16BF00A51B8A&&FORM=VRDGAR

https://www.bing.com/videos/search?q=Origins+of+Nanotechnology+video&&view=detail&mid=EDFE7EFD2228391E2798EDFE7EFD2228391E2798&&FORM=VDRVRV

Introduction:

The word Nanotechnology was first introduced in 1974 by Norio Taniguchi. Norio Taniguchi was a professor of Tokyo University of Science. He coined the term nano-technology to describe semiconductor processes such as thin film deposition and ion beam milling exhibiting characteristic control on the order of a nanometer: *"Nano-technology' mainly consists of the processing of separation, consolidation, and deformation of materials by one atom or one molecule."*

Taniguchi started his research on abrasive mechanisms of high precision machining of hard and brittle materials. At Tokyo University of Science, he went on to pioneer the application of energy beam techniques to ultra-precision materials processing. These included electro discharge, microwave, electron beam, photon (laser) and ion beams.

He studied the developments in machining techniques from 1940 until the early 1970s and predicted correctly that by the late 1980s, techniques would have evolved to a degree that dimensional accuracies of better than 100 nm would be achievable.

Since then, scientists and engineers have defined nanotechnology as a science of matter 1 billionth of a meter (10^{-9}) in size. Nanotechnology utilizes concepts from physics, chemistry, and materials science in effort to explain the unique behavior of nanoscale materials.

Recommended videos:

https://www.bing.com/videos/search?q=Origins+of+Nanotechnology+video&&view=detail&mid=749098B91A1BDA9F526A749098B91A1BDA9F526A&&FORM=VRDGAR

Everyday Nanotechnology:

Nanoscale materials (engineered nanoparticles – ENPs) are employed in a wide range of everyday products. We can find ENPs in cosmetics, personal care products, detergents, foods (processing and packaging), and clothing. Additional examples include the employment of carbon nanotubes – hollow and cylindrical – in composition of polymers, electromagnetic shielding, electron field emitters (flat panel monitors, displays), super capacitors, batteries, hydrogen storage, carbon fiber fabric, bullet proof wastes, structural compounds, and many more.

Significant advances have been achieved in the semiconductor industry due to the incorporation of nanotechnology. For example, cellphones became the devices with multiple application and often, quite complicated. This is possible today because these devices comprise of increasingly powerful computer chips with an extensive amount of nanoscale transistors influencing the functionality of cell phones. If we tried to build a phone with the same capability years ago and without nanotechnology, it would be the size of a building and probably not incorporate all applications.

Computers have become compact, more versatile, and increasingly powerful due to nanotechnology and the developments in processes and materials. Of course, the design of micro components leads the way. The first integrated circuit (IC) developed back in the late 1950s by Jack Kilby of Texas Instrument and Robert Noyce of Fairchild Semiconductor made the long way and became what they are now – microchips.

In truth, one can find traces of nanotechnology in almost everything and people used it since the ancient times without even understanding it. They created artifacts employing glass that changed colors depending on the angle of observation (Lycurgus cup). They made weapons that were stronger, more flexible, and sharper (Damascus Swords). They applied gold and silver particles to other metals and made stained glass that reflected light differently at different times of the day. That all served the intended purpose.

The application of nanotechnology in art, weaponry, and scientific research is known and understood but where did it come from. Where did people see it to understand the importance and the usefulness of it? It came from nature. Bright and vivid colors seen in nature often ascend from the collaboration of light with periodically arranged, micro- and nanoscale structures. For instance, the wings of butterflies contain a amalgamation of multilayer optical grating and other unique edifices, which produce an array of complex colors. These structures are wavelength-selective and reflect specific wavelengths over a far-reaching range of angles.

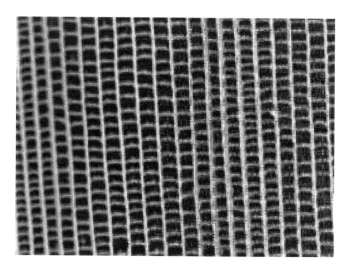

Scanning Electron Microscope image of a butterfly wing.

The nature granted many unique designs and structures to almost all animal and plant life allowing them to survive the harsh conditions and the life-threatening events. Skin, feet, fur, eyes, changing color and size, and quick adaptation are the attributes of the world surrounding us. How could we not learn?

Recommended videos:

https://www.bing.com/videos/search?q=Everyday+Nanotechnology+video&&view=detail&mid=553A7CD4964D5E20C530553A7CD4964D5E20C530&&FORM=VDRVRV

https://www.bing.com/videos/search?q=Everyday+Nanotechnology+video&&view=detail&mid=E04C27CBF9BAAF838428E04C27CBF9BAAF838428&&FORM=VDRVRV

2.0 Chemistry Foundations in Nanotechnology

Introduction:

Chemistry is an integrated part of our nature and it has been the driving force behind most of the technological advances. Materials with sizes from 0.1 to 10nm can be created shattering and creating bonds between atoms or groups of atoms (molecules). Innovative chemical reactions have led to the production of uniform nanostructures with numerous shapes (rods, spheres, cubes, wires, and anything in between) and arrangements (organics, metals, oxides, and semiconductors).

Recommended videos:

https://www.bing.com/videos/search?q=Chemistry+Foundations+in+Nanotechn ology+video&&view=detail&mid=3F9F8E2797A68D872B223F9F8E2797A68D 872B22&&FORM=VRDGAR

https://www.bing.com/videos/search?q=introduction+to+nanotechnology+video &&view=detail&mid=84925496A4B8E062FE0184925496A4B8E062FE01&&F ORM=VRDGAR

https://www.bing.com/videos/search?q=introduction+to+nanotechnology+video &qpvt=introduction+to+nanotechnology+video&view=detail&mid=4E4FD874B D7C61E63E604E4FD874BD7C61E63E60&&FORM=VRDGAR

Atoms:

The basic model of an atom consists of a dense, positively charged nucleus surrounded by negatively charged electrons orbiting the nucleus. It looks like a very nice and balanced picture. An atom is the smallest constituent unit of ordinary matter that has the properties of a chemical element and most of the time follows the laws of physics. Every solid, liquid, gas, and plasma are composed of neutral or ionized atoms. Atoms are extremely small. They are typical sizes around 100 picometers (a ten-billionth of a meter, in the short scale).

1.   Atoms are the rudimentary units of matter and the defining structure of elements. The term "atom" comes from the Greek word for indivisible (it was once thought that atoms were the smallest units in the universe and could not be divided). We now know that atoms are made up of three types of particles: protons, neutrons and electrons - which are composed of even smaller particles such as quarks. Now we understand them better

2.   Atoms were created after the Big Bang 13.7 billion years ago. As the superhot, dense new universe cooled, environments became suitable for quarks and electrons to form. Quarks came together to form protons and neutrons, and these particles combined into nuclei. This took place within the first few minutes of the universe's existence, according to **CERN**.

3.   All atoms have a specific quantity of protons - the atomic number (Periodic Table). Atoms do not have the electrical charge because the number of negatively charged electrons exactly balances the positive charge of the nucleus, thus there one electron for every proton.

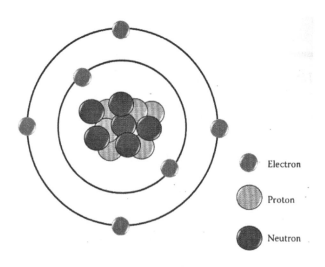

Electron

Proton

Neutron

Basic model of the atom.

Ions:

During the chemical reaction, the nucleus of an atom remains unchanged, yet electrons could be lost or gained. When atom lose or gained electrons, a changed particle is called an ion. In short, an ion is an atom or molecule that has a net electrical charge. Since the charge of the electron is equal and opposite to that of the proton, the net charge of an ion is non-zero due to its total number of electrons being unequal to its total number of protons.

A cation is a positively charged ion, with fewer electrons than protons, while an anion is negatively charged ion, with more electrons than protons. Because of their opposite electric charges, cations and anions attract each other and readily form compounds.

Largely, metal atoms lose electrons and nonmetal atoms gain electrons. Positive ions are smaller than neutral atoms because there are fewer electrons, and they are more closely held by the positive charge of the nucleus. Thus, negative ions are larger than the neutral atoms due to the extra electrons added to the electron group.

Oxidation and Reduction:

Often ions are used as the building blocks for nanomaterials. Nanomaterial synthesis frequently involves a chemical reaction that would be ferrying electrons between particles. Oxidation occurs when an atom or ion loses electrons due to any reasons: oxygen, moisture, temperature or a chemical reaction. Reduction occurs when an atom or ion gain electrons. A particle going through oxidation is called the reducing agent, and one that went through reduction is the oxidizing agent.

Recommended videos:

https://www.bing.com/videos/search?q=oxidation+and+reduction+video&qpvt=oxidation+and+reduction+video&view=detail&mid=ABF989D5038FCB27C620ABF989D5038FCB27C620&&FORM=VRDGAR

Subatomic Particles:

In the physical sciences, subatomic particles are particles that are much smaller than atoms. There are two types of subatomic particles: elementary particles, which according to current theories are not made of other particles; and composite particles that incorporate some other particles. Particle physics and nuclear physics study these particles and how they interact.

Subatomic particles include electrons, the negatively charged, almost massless **particles** that nonetheless account for most of the size of the atom, and they include the **Subatomic particle**: **Subatomic particle**, any of various self-contained units of matter or energy that are the fundamental constituents of all matter.

The particles that have a positive charge and help to form an atom's nucleus when combined with a neutron make the nucleus stable. The **proton** is one of the most stable sub-atomic particles and helps to determine the identity of an atomic element.

The 12 elementary particles of matter are **six quarks (up, charm, top, Down, Strange, Bottom) 3 electrons (electron, muon, tau) and three neutrinos (e, muon, tau)**. Four of these elementary particles would suffice in basis to build the world around us: the up and down quarks, the electron and the electron neutrino.

Recommended videos:

https://www.bing.com/videos/search?q=Subatomic+particles+video&qpvt=Subat omic+particles+video&view=detail&mid=C9382B8AADD8F5351F94C9382B8A ADD8F5351F94&&FORM=VRDGAR

https://www.bing.com/videos/search?q=Subatomic+particles+video&qpvt=Subat omic+particles+video&view=detail&mid=7B6CEB0B4B278948661C7B6CEB0B 4B278948661C&&FORM=VRDGAR

8 Most Important Subatomic Particles:

1. ELECTRON

While protons and neutrons (and their constituent quarks) give atoms their heft, it's their following of much lighter electrons that determines how atoms come together to form molecules - in short, it's electrons that give us chemistry.

2. PHOTON

The nature of light baffled scientists and philosophers since the most ancient times. Some insisted that light was a wave; others (Isaac Newton) said light was made up of particles. Albert Einstein proved that Sr. Newton was on the right track, realizing that light is "quantized," made of discrete particles (and behaved like a wave, too). Unlike electrons and quarks, photons have no "rest mass" - that is, they don't weigh anything, in the ordinary sense of the word. But photons still have energy. That energy turns out to be proportional to the frequency of the light, so that blue light (higher frequency) conveys more energy per photon than red light (lower frequency). More, photons carry more than just visible light. They convey all forms of electromagnetic radiation, including radio waves (much lower frequencies than visible light) and x-rays (much higher frequencies than visible light).

3. QUARK

Quarks are what most of the actual, familiar stuff in the universe is made of—you and me, stars and planets, rain and snow, and galaxies. Quarks are drawn to one another through the strong nuclear force, to form protons and neutrons, which make up the nuclei of atoms. In fact, due to the peculiarities of the rules of quantum mechanics, quarks can only exist within these larger, composite units. We can never see a quark on its own. They come in six "flavors": up, down, strange, charm, top, and bottom. The up and down quarks are the most stable, so it's those two, in specific, that most "stuff" is made of (the others can exist only under more exotic conditions).

4. NEUTRINO

Neutrinos are elusive, extremely lightweight particles that just hardly interact with matter at all. They zip through matter so effortlessly that, physicists wondered if they might have zero rest mass, like photons. First detected in the 1950s, it was only in the last couple of decades that physicists were able to show that neutrinos do, in fact, have a teensy amount of mass. While tiny, neutrinos are also ubiquitous; some 100 trillion neutrinos, created in the center of the Sun (the major source), pass through your body each second.

5. HIGGS BOSON

Nicknamed the "God particle" by Leon Lederman back in 1993, the Higgs Boson has become the most famous of all particles in the last few years. First hypothesized in the 1960s (by Peter Higgs as well as by several other physicists, working independently), it was finally captured at the Large Hadron Collider near Geneva in 2012. The particle had been the last piece of the so-called "Standard Model" of particle physics to be found. The model, developed in the 1960s, explains how the known forces operate, apart from gravity. The Higgs endow the other particles with mass.

6. GRAVITON

The graviton (if it exists) would be a "force carrier," like the photon. Photons "mediate" the force of electromagnetism and gravitons would do the same for gravity. (When a proton and an electron attract each other via electromagnetism, they exchange photons. Similarly, two massive objects that attract each other via gravitation ought to be exchanging gravitons.) This would be a way of explaining the gravitational force purely in terms of quantum field theories. The problem is that gravity is the weakest of the known forces, and there's no known way of building a detector that could capture the graviton. However, physicists know a fair bit about the properties that the graviton must have. For example, it's believed to be massless (like the photon), it should travel at the speed of light, and it must be a "spin-two boson," in the jargon of particle physics.

*In Quantum Mechanics and Particle Physics, **spin** is a basic form of angular momentum employed by elementary particles, composite particles (hadrons), and atomic nuclei. Spin is one of two types of angular momentum in Quantum Mechanics, the other being **orbital angular momentum**. In some ways, spin is like a vector quantity. It has a definite magnitude, and it has a "direction" (yet, quantization makes this "direction" different from the direction of an ordinary vector). All elementary particles of a given kind have the same magnitude of spin angular momentum, which is indicated by assigning the particle a spin quantum number.*

7. DARK MATTER PARTICLE

About 90 years ago, astronomers noticed that there's something funny about the way that galaxies move. It turns out that there isn't enough visible matter in galaxies to account for their detected motion. Since then, astronomers and physicists have been struggling to explain the "dark matter" thought to make up the missing mass. What is dark matter made of? One possibility is that it's made up of as-yet unknown fundamental particles, likely produced in the first moments after the big bang.

8. TACHYON

Ever since Einstein put forward the first part of his theory of relativity, known as "special relativity", we've known that nothing can move faster than light. Tachyons are hypothetical particles that always travel faster than light. Unfortunately, it doesn't mesh very well with what we know about the workings of the universe. Still, do not know everything yet. In fact, we know so little. But in the 1960s, some physicists found a loophole: As long as the particle was created above light speed and never traveled slower than light, it could theoretically exist.

Recommended videos:

https://www.bing.com/videos/search?q=8+Most+Important+Subatomic+Particles+video&&view=detail&mid=6077610062AA1A62EEEB6077610062AA1A62EEEB&&FORM=VRDGAR

https://www.bing.com/videos/search?q=8+Most+Important+Subatomic+Particles+video&&view=detail&mid=1F50E1BFB80B342C48861F50E1BFB80B342C4886&&FORM=VRDGAR

https://www.bing.com/videos/search?q=8+Most+Important+Subatomic+Particles+video&&view=detail&mid=AF1FB35B1D8D981FD0F3AF1FB35B1D8D981FD0F3&rvsmid=52866A81B360F64655A552866A81B360F64655A5&FORM=VDRVRV
https://www.bing.com/videos/search?q=7.%09DARK+MATTER+PARTICLE&&view=detail&mid=E772DAAC0517EB0F9DF6E772DAAC0517EB0F9DF6&&FORM=VRDGAR

Chemical Bonding:

Atoms join to form molecules, the smallest part of any compound. About thirty atoms bonded together would form a molecule one nanometer in size. Chemical bonds are important to nanotechnology because they can behave as hinges, bearings, or other types of nanoscale mechanical devices.

One type of bonding is ionic bonding and it could occur with metal and nonmetal atoms. Compounds with ionic bonding are brittle and have high melting points. The ions in those compounds arrange themselves in a crystal lattice structure.

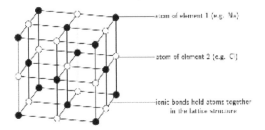

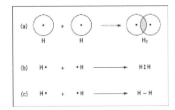

Crystal lattice structure.

Formation of a covalent bond.

Covalent bonds (second type of bonding) are formed when nonmetals share electron pairs. When two nonmetal atoms are close to each other, the positively charged nuclei repel each other and the negatively charged electrons repel each other as well. Yet, the nuclei and electrons attract each other, resulting in a concentration of negative charge between the atoms. The shared pair of electrons in any covalent bond becomes a "glue", which holds atoms together. These types of compounds tend to be gases, liquids, or solids with low melting points.

Recommended videos:

https://www.bing.com/videos/search?q=7.%09Chemical+Bonding+video&&view=detail&mid=0881BEBC05A82641903I0881BEBC05A826419031&&FORM=VRDGAR

Chemical Reaction:

A chemical reaction is a process that involves rearrangement of the molecular or ionic structure of a substance, as opposed to a change in physical form or a nuclear reaction. Chemical reactions are achieved through association of different chemical components under specific conditions.

A chemical reaction is a process that leads to the transformation of one set of chemical substances to another. Classically, chemical reactions encompass changes that only involve the positions of electrons in the forming and breaking of chemical bonds between atoms, with no change to the nuclei and can often be described by a chemical equation.

Chemical reactions are often employed to create various forms of nanomaterials and chemical equations represent the process. Those reaction could be simple or quite complicated and they always require the outmost safety procedures.

Recommended videos:

https://www.bing.com/videos/search?q=Chemical+Reaction+video&&view=detail&mid=1635AF6A7AC59A6D54241635AF6A7AC59A6D5424&&FORM=VRDGAR

https://www.bing.com/videos/search?q=Chemical+Reaction+video&&view=detail&mid=DB163DA868DEFCF2ACB0DB163DA868DEFCF2ACB0&&FORM=VRDGAR

Quantum Mechanics; Atomic and Molecular Orbitals:

Quantum Mechanics (**QM**; also known as quantum physics or quantum theory) including quantum field theory, is a fundamental branch of physics concerned with processes involving nanoscale particles, for example, atoms and photons.

Quantum Physics encompass various basic principles vis-à-vis the nature of **matter and energy**. Quantum physics is the branch of science, that deals with the behavior and characteristics of matter (in the subatomic level) and energy. It is also referred to as Quantum Mechanics.

Quantum mechanics has had **massive success in explaining many of the features of our universe**. Quantum mechanics is often the only theory that can reveal the individual behaviors of the subatomic particles that make up all forms of matter (electrons, protons, neutrons, photons, and others). Quantum theory of light says that **light is composed of tiny particles aka photons**, which exhibit wave like properties as well.

In mathematical physics and quantum mechanics, quantum logic is **a set of rules for reasoning about schemes** which takes the principles of quantum theory into account.

Rules associated with classical physics suddenly do not apply at the nanoscale. Electrons may show properties of both waves and particles and quantum mechanics was developed to describe that unique behavior. This field encompasses the probability of finding an electron in a region of space. In quantum mechanics, the properties of particles such as electrons and photons are described using wave functions (Ψ). More, the square of the wave function (Ψ^2) is proportional to the probability of finding a particle in a specific region of space. These properties are utilized in instruments employed to study the atomic scale. For instance, the scanning tunneling microscope (STM) allows users to visualize atoms due to quantum effect known as tunneling.

Atomic orbitals are derived from wave functions (Ψ) that describe the probable three-dimensional distribution of electrons. The "s" orbitals and the "p orbitals play an important role in chemical bond formation.

Molecular orbitals can be used to understand the basic operations of molecular electrons. Molecular orbitals have characteristics similar to atomic orbitals (they can hold a maximum of two electrons). Also, electrons occupying the same molecular orbital require opposite spins. Molecular orbitals have definite energies. Whenever two atomic orbitals overlap, two molecular orbitals are formed: one bonding and one antibonding.

Bonding orbitals are constructive unions of atomic orbitals and antibonding orbitals are destructive combinations of atomic orbitals. They have a nodal (intersection) plane where electron density equals zero. Molecular orbitals are involved in chemical reactions. These orbitals are referred to as the highest occupied molecular orbital **(HOMO)** and the lowest unoccupied molecular orbital **(LUMO)**. The **HOMO** electrons are the most loosely held electrons in the molecule and participate in the chemical reaction. The **LUMO** receives electrons during a chemical reaction.

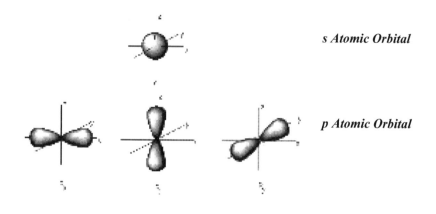

s Atomic Orbital

p Atomic Orbital

Recommended videos:

https://www.bing.com/videos/search?q=Quantum+Mechanics+video&&view=detail&mid=9106D7F211123CCAAAC39106D7F211123CCAAAC3&&FORM=VRDGAR

https://www.bing.com/videos/search?q=Atomic+and+Molecular+Orbitals+video&&view=detail&mid=50EFC5A6E03AC13DEB0B50EFC5A6E03AC13DEB0B&&FORM=VDRVRV

Intermolecular Forces:

Forces that exist between molecules are referred to as intermolecular forces. It is important to consider these forces when investigating nanotechnology. Quite often these forces are the drivers in the assembly of nanomaterials. Compared to chemical bonds, these forces require less energy to disrupt them. Encouraging or disrupting intermolecular forces is accountable for phase changes in materials, yet, when molecular substance changes phase (liquid to gas; solid to liquid), the molecules continue to stay intact. Physical properties (boiling points, melting points, etc.) reflect the strength of the intermolecular forces. There are three major types of intermolecular forces: London dispersing force; dipole-dipole interaction; and hydrogen bond.

The London dispersion force is the weakest intermolecular force. The **London dispersion force** is a temporary attractive **force** that initiates when the electrons in two adjacent atoms occupy positions that make the atoms form temporary dipoles. This **force** is sometimes called an induced, dipole-induced, and dipole attraction. **London forces** are the attractive **forces** that cause nonpolar substances to condense to liquids and to freeze into solids. These forces create the whole new array of applications that are constantly under study and consideration for applications.

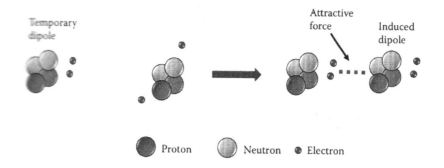

Dipole–dipole interactions are a type of intermolecular attractions between two molecules. Dipole-dipole interactions are electrostatic interactions between the permanent dipoles of different molecules. These interactions align the molecules to increase the attraction. An electric monopole is a single charge, while a dipole is two opposite charges closely spaced to each other. Molecules with dipoles are polar and are very abundant in nature. A physical **dipole** consists of two equal and opposite point charges - two poles. Its field at large distances (distances large in comparison to the separation of the poles) depends almost entirely on the **dipole** moment as defined above.

A **hydrogen bond** is an attraction between a slightly positive hydrogen on one molecule and a slightly negative atom on another molecule. Hydrogen bonds are dipole-dipole forces. The large electronegativity difference between hydrogen atoms and several other atoms, such as fluorine, oxygen and nitrogen, causes the bond between them to be polar. The other atoms have more attraction for the shared electrons, so they become slightly negatively charged and hydrogen becomes slightly positively charged. Hydrogen atoms are small, so they can get very close to other atoms. This allows them to connect with the slightly negatively charged unshared electron pair of a nearby atom and create a bond with it. A hydrogen bond is usually represented as a dotted line between the hydrogen and the unshared electron pair of the other electronegative atom.

Recommended videos:

*https://www.bing.com/videos/search?q=Intermolecular+Forces+video&&view=d
etail&mid=E0EB8BF42E1AB4B81DA8E0EB8BF42E1AB4B81DA8&&FORM
=VRDGAR*

Polymers:

Polymers are molecules that consist of a long, repeating chain of smaller units called monomers. They have the highest molecular weight among any molecules and may consist of billions of atoms. Human DNA is a polymer with over 20 billion constituent atoms. Proteins made up of amino acids, and many other molecules that make up life are polymers. They are long chain molecules often used in nanomaterial synthesis as precursors, templates or directional negotiators. Polymers consist of long chains of atoms attached together through covalent bonds. Polymers can be made stiffer by utilizing chemical bonds between chains. This process is referred to as cross-linking. The greater the number of cross-links, the more rigid the polymer is.

Adjacent polymer chains are held together by weak intermolecular forces. Polymers are molecules formed via ***polymerization*** (joining together) of smaller molecules referred to as monomers. Polymers are commonly found in abundance in nature, in plants and animals. Also, polymers are often referred as plastics and can be shaped using heat and pressure. **Thermoplastics** are particles that can be reshaped with heat and pressure. They also could be melted down and recycled for other purposes. A **thermoset** is a polymetric material shaped using a chemical process and cannot be reshaped at will. **Elastomers** are polymers exhibiting a rubbery or elastic performance. When these materials are stretched or bent, they return to the original shape as soon as the force is removed.

Polymer, any of a class of natural or synthetic substances composed of very large molecules, called macromolecules, that are multiples of simpler chemical units called monomers. Polymers make up many of the materials in living organisms, including, for instance, proteins, cellulose, and nucleic acids. Moreover, they establish the basis of such minerals as diamond, quartz, and feldspar and such man-made materials as concrete, glass, paper, plastics, and rubbers.

Recommended videos:

https://www.bing.com/videos/search?q=Polymers+video&&view=detail&mid=47 D5B56BF7A6465D6D4447D5B56BF7A6465D6D44&&FORM=VRDGAR

https://www.bing.com/videos/search?q=Polymers+video&&view=detail&mid=E7 B885ED19887444E512E7B885ED19887444E512&&FORM=VRDGAR

Semiconductor Materials:

Semiconductor materials have both metallic and nonmetallic properties Yet, intentionally introduced impurities to levels as low as 0.01% can change the electrical resistance of a semiconductor 10,000-fold. Switching between the metallic and insulating characteristics of semiconductors is the basis of the transistor, an electronic switch in modern electronics. Semiconductors contain electrons and holes responsible for carrying charges through semiconductor materials. The type of charge carriers in the semiconductor vary through controlling the type concentration of dopants added to the material.

Dopant - any impurity deliberately added to a semiconductor to modify its electrical conductivity. The most commonly employed elemental semiconductors are silicon and germanium, which form crystalline lattices where each atom shares one electron with each of its four nearest neighbors. If a small proportion of the atoms in such a lattice is swapped by atoms such as phosphorus or arsenic, where five electrons are available for bond formation, the extra electron of such dopant atom becomes available for electrical conduction. The semiconductor is then said to be doped with phosphorus or arsenic that are called donor atoms, and the semiconductor is classed as *n*-type (*n* for negative, because the charge carriers are electrons - negatively charged particles). Doping with atoms such as boron or indium that have only three electrons available, creates a positively charged site, or "**hole**," in the bonding arrangement. Conduction can occur by migration of the positively charged site through the crystal lattice, and a semiconductor doped with an atom of this type, an acceptor atom, is called *p*-**type**.

Semiconductor, any of a class of crystalline solids intermediate in electrical conductivity between a conductor and an insulator. Semiconductors are employed in the manufacture of various kinds of electronic devices, including diodes, transistors, and integrated circuits. Transistors, diodes and integrated circuits can all be classified as semiconductor devices because they are made from semiconductor materials.

Recommended videos:

https://www.bing.com/videos/search?q=Semiconductor+Materials+video&&view=detail&mid=48EDFC92A9B89246907148EDFC92A9B892469071&&FORM=VRDGAR

https://www.bing.com/videos/search?q=semiconductor+Dopant+video&&view=detail&mid=049EC7F5D2C2FF89F442049EC7F5D2C2FF89F442&&FORM=VRDGAR

3.0 Physics and Nanotechnology

Introduction:

Nanotechnology is a highly multidisciplinary field, drawing from fields such as **applied physics, materials science, colloidal science, device physics, supramolecular chemistry**, and even mechanical and electrical engineering.

Nanotechnology is a field of applied sciences and technologies involving the control of matter on the atomic and molecular scale, normally below 100 nanometers. Nanomaterials may exhibit different physical and chemical properties compared with the same substances at normal scale, such as increased chemical reactivity due to greater surface area.

There are unique light-matter interactions occurring at the nanoscale. Nanoparticles with diameter less than 100nm in size can be converted to absorb some specific wavelengths of light while reflecting other wavelengths. Even smaller nanoparticles, known as quantum dots, with diameters between 1 and 10nm, can be converted to emanate some specific wavelengths of light when exposed to ultraviolet radiation.

The technical definition of nanotechnology is the **manipulation of atoms and molecules to produce a more significant product, which essentially makes it an amalgamation of physics, chemistry and biology**. In layman's term, nanotechnology is the study of atoms and how they interact.

Recommended videos:

https://www.bing.com/videos/search?q=Physics+and+Nanotechnology+video&&view=detail&mid=8CB8F662EB19086703938CB8F662EB1908670393&&FORM=VRDGAR

Electromagnetic Radiation:

Electromagnetic radiation is energy traveling through space in the form of waves. It is a kind of radiation that includes visible light, radio waves, gamma rays, and X-rays, in which electric and magnetic fields vary simultaneously. Electromagnetic radiation, in classical physics, the flow of energy at the universal speed of light through free space or through a material medium in the form of the electric and magnetic fields that make up **electromagnetic** waves such as radio waves, visible light, and gamma **rays**.

Electromagnetic radiation consists of **electromagnetic waves**, which are synchronized oscillations of electric and magnetic fields that circulate at the speed of light, which, in a vacuum, is commonly denoted c. The wave front of electromagnetic waves emitted from a point source (such as a light bulb) is a sphere. The position of an electromagnetic wave within the electromagnetic spectrum can be characterized by either its frequency of oscillation or its wavelength. Electromagnetic waves of different frequency are called by different names since they have different sources and effects on matter. In order of increasing frequency and decreasing wavelength these are: radio waves, microwaves, infrared radiation, visible light, ultraviolet radiation, X-rays and gamma rays.

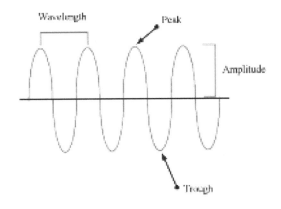

Electromagnetic wave.

Electromagnetic Radiation consists of oscillating, perpendicular electric and magnetic fields.

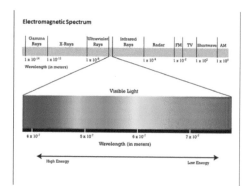

All forms of Electromagnetic Radiation are listed on the Electromagnetic spectrum.

All forms of electromagnetic radiation travel in a vacuum at the speed of light, *c*, with a value of 3.0 x 10^8 m/s.

Recommended videos:

https://www.bing.com/videos/search?q=Electromagnetic+Radiation+video&&view=detail&mid=1467D9B45E94F3E4344E1467D9B45E94F3E4344E&&FORM=VRDGAR

https://www.bing.com/videos/search?q=Electromagnetic+Radiation+video&&view=detail&mid=786580BD4E639E76A39F786580BD4E639E76A39F&&FORM=VRDGAR

The Wave Nature of Light:

Light is a **transverse, electromagnetic wave** that can be seen by humans. The wave nature of light was first illustrated through experiments on diffraction and interference. Like all electromagnetic waves, light can travel through a vacuum.

In an **electromagnetic wave**, electric and magnetic field vectors are perpendicular to each other and at the same time are perpendicular to the direction of propagation of **wave**. This **nature** of **electromagnetic wave** is known as Transverse **nature**.

Recommended videos:

https://www.youtube.com/watch?v=Io-HXZTepH4

https://www.bing.com/videos/search?q=The+Wave+Nature+of+Light+video&&view=detail&mid=EDC45520EC9D51028956EDC45520EC9D51028956&&FORM=VRDGAR

Nature of Light
Wave Properties

- Light is a self-propagating electro-magnetic wave
  - A time-varying electric field makes a magnetic field
  - A time-varying magnetic field makes an electric field

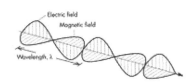

- Wavelength (or frequency) are related to energy
- Wave amplitude → brightness
- Angle of field lines → polarization

Photoelectric Effect:

The photoelectric effect is the observation that many metals emit electrons when light shines upon them. Electrons emitted in this manner can be called photoelectrons. The phenomenon is commonly studied in electronic physics, as well as in fields of chemistry, such as quantum chemistry or electrochemistry.

The **photoelectric effect** is the emission of electrons or other free carriers when light hits a material. Electrons emitted in this manner can be called *photoelectrons*. This phenomenon is commonly studied in electronic physics, as well as in chemistry, such as quantum chemistry and electrochemistry.

Photoelectric effect, phenomenon in which electrically charged particles are released from or within a material when it absorbs electromagnetic radiation. The effect is often defined as the ejection of electrons from a metal plate when light falls on it. In a broader definition, the radiant energy may be infrared, visible, or ultraviolet light, X rays, or gamma rays; the material may be a solid, liquid, or gas; and the released particles may be ions (electrically charged atoms or molecules) as well as electrons. The phenomenon was fundamentally significant in the development of modern physics because of the puzzling questions it raised about the nature of light - particle versus wavelike behavior - that were finally resolved by Albert Einstein in 1905. The effect remains important for research in areas from materials science to astrophysics, as well as forming the basis for a variety of useful devices.

Photoelectric Effect

- Energy In > Energy Out   What absorbed the missing energy?

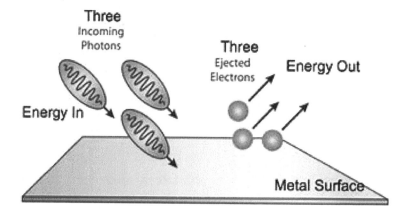

Recommended videos:

https://www.bing.com/videos/search?q=Photoelectric+Effect+video&&view=detail&mid=0281E7C555F3934C818C0281E7C555F3934C818C&&FORM=VRDGAR

https://www.bing.com/videos/search?q=Photoelectric+Effect+video&&view=detail&mid=B7487C21E32B5B55A870B7487C21E32B5B55A870&&FORM=VRDGAR

Band Structure:

In solid-state physics, the electronic band structure (or simply band structure) of a solid describes the range of energies that an electron within the solid may have (called energy bands, allowed bands, or simply bands) and ranges of energy that it may not have (called band gaps or forbidden bands). **Band theory** or **band structure** describes the quantum-mechanical behavior of electrons in solids. Inside isolated atoms, electrons possess only certain discrete energies, which can be depicted in an energy-level diagram as a series of distinct lines.

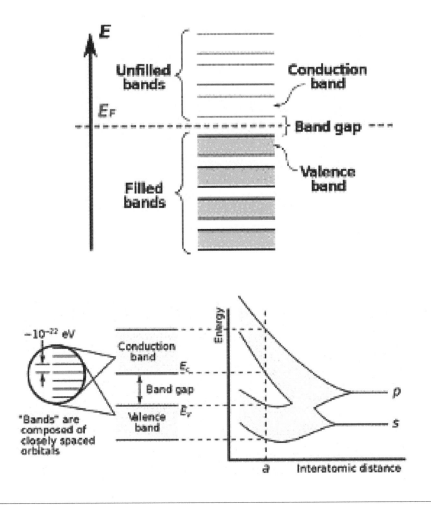

Recommended videos:

https://www.bing.com/videos/search?q=Band+Structure+video&&view=detail&mid=D36622D2579AB8E0BF43D36622D2579AB8E0BF43&&FORM=VRDGAR

Band Diagram:

In solid-state physics of semiconductors, a band diagram is a diagram plotting various key electron energy levels (Fermi level and nearby energy band edges) as a function of some spatial dimension, which is often denoted x. These diagrams help to explain the operation of many kinds of semiconductor devices and to visualize band bending. Orbital arrangements could be visualized employing a band diagram.

The low energy band, the valence band, consists of orbitals containing electrons. The higher energy band consists of unfilled orbitals and is referred to as the conduction band. Depending on the nature of the material investigated, there can be a space between the valence band and the conduction band that is known as the band gap. Nonconducting materials or insulators have a wide band gap, decreasing the probability of electrons transfer from the valence band to the conduction band. Insulators are poor conductors of electricity due to the large band gap.

Recommended videos:

https://www.bing.com/videos/search?q=Band+Diagrams+Chemistry&&view=detail&mid=D9C10BD3051F98E319B4D9C10BD3051F98E319B4&&FORM=VDRVRV

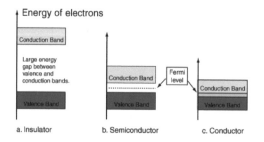

Classification of Energy Bands:
a) Insulators
b) Semiconductors
c) Conductors

Conductors:

Gold, Aluminum, Silver, Copper, metals, allow electric current to flow through them. There could be an energy overlap as valence band is fully filled and conduction band is partially filled.

Recommended videos:

https://www.bing.com/videos/search?q=Conductors+video&&view=detail&mid=A633F6FD09CC84B39B06A633F6FD09CC84B39B06&&FORM=VRDGAR

Insulators:

Glass and wood are the examples of the insulator; these substances do not allow electricity to pass through them. They have high resistivity and very low conductivity.

Recommended videos:

https://www.bing.com/videos/search?q=Insulators+video&&view=detail&mid=CD2A5ED0A1F187E17AE6CD2A5ED0A1F187E17AE6&&FORM=VRDGAR

Semiconductors:

Germanium and Silicon are the most preferable material whose electrical properties lie in between semiconductors and insulators. The energy band diagram of semiconductor is shown where the conduction band is empty, and the valence band is filled but the forbidden gap between the two bands is very small. Thus, semiconductor requires small conductivity.

Recommended videos:

https://www.bing.com/videos/search?q=Semiconductors+video&&view=detail&mid=55AD70BCC2E1DAB57ADB55AD70BCC2E1DAB57ADB&&FORM=VRDGAR

Fermi Level:

Free electrons are present in all materials, but the numbers of free electrons differ from material to material. There are fewer free electrons present in insulators, while free electrons are present in abundance in conductors, and semiconductors have a moderate number of free electrons.

The Fermi level is the total chemical potential for electrons (or electrochemical potential for electrons) and is usually denoted by μ or **EF**. The Fermi level of a body is a thermodynamic quantity, and its significance is the thermodynamic work required to add one electron to the body (not counting the work required to remove the electron from wherever it came from). In short, it's an amount of energy required to add an electron.

For semiconductors, the Fermi level usually lies almost at the **middle of the band gap**. The level has equal probability of occupancy for the electrons as well as holes. Fermi Level or Fermi Energy is a **quantum phenomenon**, which translates as the difference in energy state occupied by the lowest level (close to the nucleus) electrons to the highest level (away from the nucleus) electrons, it can also be measured at the top of the collection of electron energy levels at absolute zero temperature.

The **Fermi level** is like the surface of that sea at absolute zero where no electrons will have enough energy to rise above the surface. The concept of the **Fermi energy** is a crucially important concept for the understanding of the electrical and thermal properties of solids but that is in another course.

Recommended videos:

https://www.bing.com/videos/search?q=Fermi+Level+video&&view=detail&mid=F4A788DCAD833A6C9B78F4A788DCAD833A6C9B78&&FORM=VRDGAR

https://www.bing.com/videos/search?q=Fermi+Level+video&&view=detail&mid=B09B79E0280C7832D232B09B79E0280C7832D232&&FORM=VRDGAR

The Bohr-Exciton Radius:

When a material absorbs a photon of sufficient energy, electrons are prompted from the valence (number of electrons available for chemical reaction) band to the conduction band, creating an electron-hole pair. This electron-hole pair is called an exciton.

A Bohr-Exciton Radius is the **distance in an electron-hole pair**. A **Quantum Dot** is a semiconductor so small that the size of the crystal is on the same order as the size of the Bohr-Exciton Radius. This unique size property causes the "band" of energies to turn into discrete energy levels.

Recommended videos:

https://www.bing.com/videos/search?q=Bohr+Radius+Explained&&view=detail&mid=F2141035D943218EBE8CF2141035D943218EBE8C&&FORM=VDRVRV

4.0   Allotropic Carbon-Based Nanomaterials

Introduction:

The word *carbon* is based on the Latin word *carbo* meaning charcoal. There are many forms of pure carbon with nanoscale dimensions. **Carbon** is a chemical element with the symbol **C** and atomic number 6. It is nonmetallic and tetravalent (valence of 4) - making four electrons available to form covalent chemical bonds. It belongs to group 14 of the periodic table (see the periodic table). Carbon is one of the few elements known and widely used since antiquity.

Recommended videos:

https://www.bing.com/videos/search?q=Carbon+video&&view=detail&mid=8FB 72204665AB8770A488FB72204665AB8770A48&&FORM=VRDGAR

https://www.bing.com/videos/search?q=Carbon+video&&view=detail&mid=ED B74E84AD86784B51F3EDB74E84AD86784B51F3&&FORM=VDRVRV

Carbon Allotropes:

Before examining nanoscale forms of carbon, it is important to discuss the most common forms of this element. Naturally occurring carbon-based materials exist in the form of allotropes, multiple forms of the same element with different atomic structures. **Carbon** can form many **allotropes** due to its valency and abundance of Carbon in nature. Well-known forms of **carbon** include diamond, coal and graphite. In recent decades many more **allotropes** and forms of **carbon** have been discovered and researched including ball shapes such as buckminsterfullerene and sheets such as graphene. Larger scale structures of **carbon** include nanotubes, nanobuds and nanoribbons and then, there is Carbone 60 (C_{60}). Carbon is a friend and an enemy of the human beings at the same time.

ALLOTROPIC FORMS OF CARBON

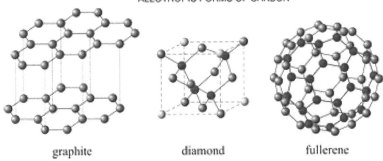

| graphite | diamond | fullerene |

Allotropy or allotropism is the property of some chemical elements to exist in two or more different forms, in the same physical state, known as allotropes of these elements. Allotropes are different structural modifications of an element; the atoms of the element are bonded together in a different manner.

Recommended videos:

https://www.bing.com/videos/search?q=Allotropic+Carbon-Based+Nanomaterials+video&&view=detail&mid=75B948196D8411C7815D75B948196D8411C7815D&&FORM=VRDGAR

https://www.bing.com/videos/search?q=Carbon+Allotropes+video&&view=detail&mid=68C9A4E25E09B213BB4968C9A4E25E09B213BB49&&FORM=VRDGAR

C_{60}:

The discovery of C_{60} (carbon in the late 1980s, had opened up new avenues in carbon research. This is a very exciting field for research and development of carbon-based materials. C_{60} is a spherical carbon molecule with 60 carbon atoms and looks like a soccer football. It contains 12 five-member rings isolated by 20 six-membered rings. Each hexagon has three pentagons and three hexagons as its neighbors, thus pentagons are surrounded by hexagons. This structure provides C_{60} molecule with extraordinary stability. Carbon atoms found in C_{60} have four valence electrons available for bonding.

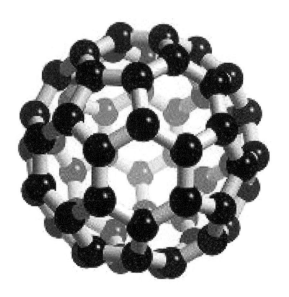

Recommended videos:

*https://www.bing.com/videos/search?q=C60+video&&view=detail&mid=C05BF
9C17F1A95C48589C05BF9C17F1A95C48589&&FORM=VRDGAR*

Endohedral and Exohedral Modifications:

Due to cage-like structure of C_{60}, they the potential for use as molecular storage for a variety of substances. Employing C_{60} as a *nanocontainer* is a good example of endohedral modification, which incorporates individual atoms, molecules, or ions within the C_{60} cage.

Endohedral fullerenes, also called endofullerenes, are fullerenes that have additional atoms, ions, or clusters enclosed within their inner spheres. The first lanthanum C_{60} complex was synthesized in 1985 and called $La@C_{60}$. The @ (at sign) in the name reflects the notion of a small molecule trapped inside a shell.

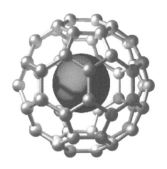

C_{60} molecules can rapidly bond with atoms of other materials, which makes it possible to modify the outside of the C_{60} cage. Attachment of atoms or molecules to the surface of the C_{60} cage is referred to as exohedral modification.

Metal - doped C_{60} exhibits superconducting characteristics (first discovered in 1991). These superconducting structures are created when C_{60} molecules assemble to form face-centered cubic (FCC) crystal structure held together via van der Waals forces. The C_{60} crystal is an insulator. Nonetheless, when atoms of alkali metals are introduced to the crystal structure, C_{60} is converted to a superconductor.

In physical chemistry, the van der Waals forces (or van der Waals' interaction), named after Dutch scientist Johannes Diderik van der Waals, are the residual attractive or repulsive forces between molecules or atomic groups that do not arise from a covalent bond, or electrostatic interaction of ions or of ionic groups with one another or with neutral molecules.

A covalent bond is a chemical bond that involves the sharing of electron pairs between atoms. These electron pairs are known as shared pairs or bonding pairs, and the stable balance of attractive and repulsive forces between atoms, when they share electrons, is known as covalent bonding.

Recommended videos:

https://www.bing.com/videos/search?q=Endohedral+video&&view=detail&mid=DD3CC9300BF0276B3075DD3CC9300BF0276B3075&&FORM=VRDGAR

https://www.bing.com/videos/search?q=Endohedral+video&&view=detail&mid=3AD650637E6C35062C4F3AD650637E6C35062C4F&&FORM=VDRVRV

CNTs:

Carbon nanotubes are a class of nanomaterials that consist of a two-dimensional hexagonal lattice of carbon atoms, bent and joined in one direction to form a hollow cylinder. Carbon nanotubes are one of the allotropes of carbon, specifically a class of fullerenes, intermediate between the buckyballs and graphene.

Carbon nanotubes (CNTs) are an allotrope of carbon. They take the form of cylindrical carbon molecules and have novel properties that make them potentially useful in a wide variety of applications in nanotechnology, electronics, optics and other fields of materials science. They exhibit extraordinary strength and unique electrical properties and are efficient conductors of heat. Inorganic nanotubes have also been synthesized. Nanotubes are members of the fullerene structural family, which also includes buckyballs. Whereas buckyballs are spherical in shape, a nanotube is cylindrical, with at least one end typically capped with a hemisphere of the buckyball structure.

Their name is derived from their size, since the diameter of a nanotube is on the order of a few nanometers (approximately 50,000 times smaller than the width of a human hair), while they can be up to several millimeters in length. There are two main types of nanotubes: single-walled nanotubes (SWNTs) and multi-walled nanotubes (MWNTs).

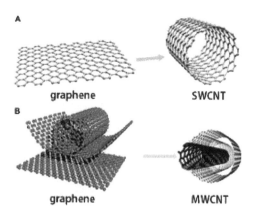

A nanotube consists of one or more seamless cylindrical shells of graphitic sheets (as shown above). Each shell is made of a hexagonal grouping of carbon atoms establishing a system connected without any edges. Nanotubes can possess either metallic or semiconducting properties depending on helicity (a combination of the spin and the linear motion of a subatomic particle) and diameter. The diameter of a nanotube also affects the mechanical characteristics of CNTs. It can impact applications ranging from scanning probe microscopy tips, electromechanical devices, and structural amalgams. The mechanical deformations can affect the electrical properties of the nanotubes.

This is still a theory, but experiments demonstrate some truth to it. CNTs have a diameter ranging from 1 to 20nm and even a very small deformation could mean a lot. The length of the CNTs could be anywhere from being 50,000 times smaller than the width of a human hair, and up to 18 centimeters.

CNTs can consist of a single tube - single-walled CNTs or SWCNTs or can contain several concentric tubes – multiwalled CNTs or MWCNTs. Both SWCNT or MWCNT can easily bundle together to form ropes due to attractive London dispersion forces, the same forces that hold sheets of graphite together. In short, CNTs are single sheets of graphite wrapped in a cylindrical form. Controlling the direction of the graphite wrapping, we could control the helicity, thus some of the mechanical characteristics of the CNTs. There are three types of CNTs (nanotube variations) wrapping: armchair, zigzag, and chiral.

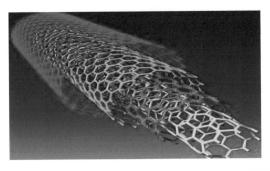

Carbon Nanotube

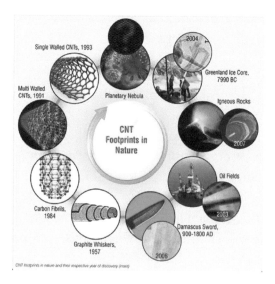

$C_a = n1.\overline{a1} + n2.\overline{a2}$

Wrapping vector

Electrical conductivity:

1. Metallic (when |n1 n2| is a multiple of 3) with zero bandgap
2. Semiconducting with finite bandgap.

armchair   zigzag   chiral

Chiral - asymmetric in such a way that the structure and its mirror image are not superimposable. Chiral compounds are typically optically active; large organic molecules often have one or more chiral centers where four different groups are attached to a carbon atom.

Recommended videos:

https://www.bing.com/videos/search?q=Carbone+Nanotubes+video&qpvt=Carbone+Nanotubes+video&view=detail&mid=9E6AB38ADDAF8CA0E93F9E6AB38ADDAF8CA0E93F&&FORM=VRDGAR

https://www.bing.com/videos/search?q=Carbone+Nanotubes+video&qpvt=Carbone+Nanotubes+video&view=detail&mid=CCF194D191335E531F00CCF194D191335E531F00&&FORM=VRDGAR

Mechanical Properties of CNTs:

The **mechanical properties of carbon nanotubes** reveal them as one of the strongest materials in nature. Carbon nanotubes (CNTs) are somewhat long hollow cylinders of graphene. Although graphene sheets have 2D symmetry, carbon nanotubes have different properties in axial and radial directions. It has been shown that CNTs are especially strong in the axial direction.

Carbon nanotubes are the strongest and stiffest materials yet discovered in terms of tensile_strength and elastic_modulus respectively. This strength results from the covalent bonds formed between the individual carbon atoms. Under excessive tensile strain, the tubes will undergo plastic deformation, which means the deformation is permanent. This deformation begins at strains of approximately 5% and can increase the maximum strain the tubes undergo before fracture by releasing strain energy.

CNTs are not nearly as strong under compression. Because of their hollow structure and high aspect ratio, they tend to undergo buckling when placed under compressive, torsional, or bending stress. Interesting that applying a force on the tip of the nanotube will cause it to bend without breaking. And, removal of the force will allow the tube to recover its original state. More, under the application of large loads, CNTs bend at extremely large angles.

Recommended videos:

https://slideplayer.com/slide/8020639/

Radial elasticity:

On the other hand, there was evidence that in the radial direction CNTs are rather soft. The first transmission electron microscope observation of radial elasticity suggested that even the van der Waals forces can deform two adjacent nanotubes. Later, nanoindentations with atomic force microscope were performed by several groups to quantitatively measure radial elasticity of multiwalled carbon nanotubes and tapping/contact mode atomic force microscopy was also performed on single-walled carbon nanotubes. Young's modulus showed that CNTs are in fact very soft in the radial direction.

Radial direction elasticity of CNTs is important especially for carbon nanotube composites where the embedded tubes are subjected to large deformation in the transverse direction under the applied load on the composite structure. One of the main problems in characterizing the radial elasticity of CNTs is the knowledge about the internal radius of the CNT; carbon nanotubes with identical outer diameter may have different internal diameter (or the number of walls). Still, they can bend, and we can predict the results.

Recommended videos:

https://www.bing.com/videos/search?q=CNTs+properties+video&&view=detail&mid=CDB3AF66D67C0C8E9C5DCDB3AF66D67C0C8E9C5D&&FORM=VRDGAR

Hardness:

Standard single-walled carbon nanotubes can withstand a pressure up to 25 GPa (gigapascal) without [plastic/permanent] deformation. They then undergo a transformation to super hard phase nanotubes. Maximum pressures measured using current experimental techniques are around 55 GPa. However, these new super hard phase nanotubes collapse at an even higher, although unknown, pressure.

The bulk modulus of super hard phase nanotubes is 462 to 546 GPa, even higher than that of diamond (420 GPa for single diamond crystal). All that would provide us with even more opportunities for new materials.

Recommended videos:

https://www.bing.com/videos/search?q=CNTs+Hardness+video&&view=detail&mid=82AB905646F1FD97797882AB905646F1FD977978&&FORM=VRDGAR

Wettability:

Wetting is the ability of a liquid to maintain contact with a solid surface, resulting from intermolecular interactions when the two are brought together. The degree of wetting (wettability) is determined by a force balance between adhesive and cohesive forces. Wetting deals with the three phases of materials: gas, liquid, and solid.

Wettability (the quality or state of being wettable: the degree to which something can be wet) is a fundamental property of a solid surface, which plays important roles in daily life, industry, and agriculture. The **wettability** of solid substrates is governed by their surface free energy and surface geometrical structure. In nature, many biological materials exhibit excellent surface **wettability**. The surface wettability of CNTs is of importance for its applications in various settings. Although the intrinsic contact angle of graphite is around 90°, the contact angles of most as-synthesized CNTs arrays are over 160°, exhibiting a superhydrophobic property. By applying a voltage as low as 1.3V, the extreme water repellant surface can be switched to a super hydrophilic one.

Recommended videos:

https://www.bing.com/videos/search?q=Wettability+video&&view=detail&mid=D93DE1E0185684D2FCDCD93DE1E0185684D2FCDC&&FORM=VRDGAR

https://www.bing.com/videos/search?q=Wettability+video&&view=detail&mid=CD56E170BDD2475D37A4CD56E170BDD2475D37A4&&FORM=VRDGAR

https://www.bing.com/videos/search?q=Wettability+video&&view=detail&mid=DE0034297E28F4D9027ADE0034297E28F4D9027A&&FORM=VRDGAR

Kinetic properties:

Multi-walled nanotubes are multiple concentric nanotubes precisely nested within one another. These exhibit a striking telescoping property whereby an inner nanotube core may slide, almost without friction, within its outer nanotube shell, thus creating an atomically perfect linear or rotational bearing. This is one of the first true examples of molecular nanotechnology, the precise positioning of atoms to create useful machines. Already, this property has been utilized to create the world's smallest rotational motor and there is no limit insight. Future applications such as a gigahertz mechanical oscillator are also envisioned. Nanoscale robots is another example of future developments. Life science is a new field where new materials and electromechanical machines can make the difference.

Recommended videos:

https://www.bing.com/videos/search?q=Kinetic+properties+video&&view=detail&mid=68ACC9F2C13DB6D51D2B68ACC9F2C13DB6D51D2B&&FORM=VRDGAR

https://www.bing.com/videos/search?q=Kinetic+properties+video&&view=detail&mid=CEF81D1D51EDEDABB7DFCEF81D1D51EDEDABB7DF&&FORM=VDRVRV

Defects:

As with any material, the existence of a crystallographic defects affects the material properties. Defects can occur in the form of atomic vacancies. High levels of such defects can lower the tensile strength by up to 85%. An important example is the Stone Wales defect, which creates a pentagon and heptagon pair by rearrangement of the bonds. Because of the very small structure of CNTs, the tensile strength of the tube is dependent on its weakest segment in a similar manner to a chain, where the strength of the weakest link becomes the maximum strength of the chain.

A Stone-Wales defect is a crystallographic defect that involves the change of connectivity of two **π-bonded** carbon atoms, leading to their rotation by 90° with respect to the midpoint of their bond. The reaction commonly involves conversion between a naphthalene - like structure into a fulvalene - like structure, that is, two rings that share an edge vs two separate rings that have vertices bonded to each other.

Recommended videos:

https://www.bing.com/videos/search?q=carbon+nanotubes+Defects+video&&view=detail&mid=EC8A72D8D3381F0C92CEEC8A72D8D3381F0C92CE&&FORM=VRDGAR

Electrical Properties of CNTs:

The unique electrical properties of CNTs are to a large extent derived from their **1-D character** and the peculiar electronic structure of graphite. Resistance occurs when an electron collides with some defect in the crystal structure of the material through which it is passing. Another interesting **property of carbon nanotube** is that their electrical resistance changes significantly when other molecules attach themselves to their carbon atoms. Researchers are using this **property** to develop sensors that can detect chemical vapors such as carbon monoxide or biological molecules. It's been noticed that CNTs can carry high currents with little to no heating.

There has been considerable practical interest in the conductivity of CNTs. CNTs with specific combinations of N and M, the structural parameters indicating how much the nanotube is twisted, can be highly conducting, and hence can be said to be metallic. Their conductivity has been shown to be a function of their chirality, the degree of twist as well as their diameter. CNTs can be either metallic or semi-conducting in their electrical behavior. Conductivity in MWNTs is quite complex. The conductivity of metallic CNTs can be more than 50 times higher than copper.

Armchair CNTs have metallic properties; chiral and zig-zag nanotubes have properties of semiconductors. The **superior properties of CNTs** are not limited to electrical and **thermal conductivities**, but also include **mechanical properties**, such as stiffness, toughness, and strength. These properties lead to a wealth of applications exploiting them, including advanced composites requiring high values of one or more of these properties.

Recommended videos:

https://www.bing.com/videos/search?q=CNTs+Electrical+Properties+video&&view=detail&mid=E8F74F642B5EB4F77811E8F74F642B5EB4F77811&rvsmid=0FCC35486CFA1CD045DF0FCC35486CFA1CD045DF&FORM=VDRVRV

CNT Electronics:

Transistors:

A transistor is a semiconductor device used to amplify or switch electronic signals and electrical power. It is composed of semiconductor material usually with at least three terminals for connection to an external circuit. A voltage or current applied to one pair of the transistor's terminals controls the current through another pair of terminals.

If cells are the building blocks of animal and plant life, **transistors** are the building blocks of the digital insurgency. Without **transistors**, the technological phenomena you use every day - cell phones, computers, cars, spaceships, and much more - would be vastly different, if they would exist at all. Before **transistors**, product engineers used vacuum tubes and electromechanical switches to complete electrical circuits and that was bulky and not very productive. Now, the products get smaller, faster, more powerful, and better.

A carbon nanotube field-effect transistor refers to a field-effect transistor that utilizes a single carbon nanotube or an array of carbon nanotubes as the channel material instead of bulk silicon in the traditional MOSFET structure. First demonstrated in 1998, there have been major developments in CNTFETs since.

Carbon nanotubes (CNTs) are perhaps the best available material for realizing nano and molecular scale electronics and sensor devices. Experiments demonstrating the use of single-wall **nanotubes** (SWNTs) as the active channel in a semiconductor (MOS) field effect **transistor** (FET) have opened the possibility for a wide range of integrated **carbon nano-tube** nanoelectronics.

Recommended videos:

https://www.bing.com/videos/search?q=Transistors+video&&view=detail&mid= 14B89840AAD43053B80514B89840AAD43053B805&&FORM=VRDGAR

https://www.bing.com/videos/search?q=Transistors+video&&view=detail&mid= E92E483DF5BDB522AE99E92E483DF5BDB522AE99&&FORM=VDRVRV

MOSFETs:

Metal oxide semiconductor field effect transistors (MOSFET) are the most commonly used transistors. It has an insulated gate; whose voltage determines the conductivity of the device. This ability to change conductivity with the amount of applied voltage can be used for amplifying or switching electronic signals. A metal-insulator-semiconductor field-effect transistor or MISFET is a term almost synonymous with MOSFET. Another synonym is IGFET for insulated-gate field-effect transistor.

The main advantage of a MOSFET is that it requires almost no input current to control the load current, when compared with bipolar transistors (bipolar junction transistors/BJTs). In an *enhancement mode* MOSFET, voltage applied to the gate terminal increases the conductivity of the device. In *depletion mode* transistors, voltage applied at the gate reduces the conductivity.

The "metal" in the name MOSFET is sometimes a contradiction, because the gate material can be a layer of polysilicon (polycrystalline silicon). Similarly, "oxide" in the name can also be a contradiction, as different dielectric materials are used with the goal of obtaining strong channels with smaller applied voltages.

The MOSFET is by far the most common transistor in digital circuits, as billions may be included in a memory chip or microprocessor. Since MOSFETs can be made with all types of semiconductors, complementary pairs of MOS transistors can be used to make switching circuits with very low power consumption, in the form of CMOS logic.

Recommended videos:

https://www.bing.com/videos/search?q=MOSFETs+video&&view=detail&mid=8DF42BA5D1C03748CF358DF42BA5D1C03748CF35&&FORM=VDRVRV

https://www.bing.com/videos/search?q=MOSFETs+video&&view=detail&mid=D672D0923A7DE3CAF6D9D672D0923A7DE3CAF6D9&&FORM=VDRVRV

CNTFETs:

A carbon nanotube field-effect transistor refers to a field-effect transistor that utilizes a single carbon nanotube or an array of carbon nanotubes as the channel material instead of bulk silicon in the traditional MOSFET structure. First demonstrated in 1998, there have been major developments in CNTFETs since.

Finally, the top gate contact is deposited on the gate dielectric, completing the process. Arrays of top-gated CNTFETs can be fabricated on the same wafer, since the gate contacts are electrically isolated from each other, unlike in the back-gated case. Also, due to the thinness of the gate dielectric, a larger electric field can be generated with respect to the nanotube using a lower gate voltage. **CNTFETs** are the most suitable alternative for the conventional

CMOS. CNTFET based circuits are expected to be 3× faster than silicon transistors, while consuming the same power. CNTFET technology has a bright future in the area of Mobile and wireless RF applications.

Recommended videos:

https://www.bing.com/videos/search?q=CNTFETs+video&&view=detail&mid=0B6836A935DF7CFA62880B6836A935DF7CFA6288&rvsmid=6324EEBD9FE194C29B786324EEBD9FE194C29B78&FORM=VDRVRV

CNT Synthesis:

Recommended videos:

https://www.bing.com/videos/search?q=Arc+Discharge+Synthesis+Video&&view=detail&mid=CD95630FEC0B8F35838CCD95630FEC0B8F35838C&&FORM=VRDGAR

Arc Discharge Synthesis:

A method for the **synthesis** of carbon nanotubes where a direct-current **arc voltage** is applied across two graphite electrodes immersed in an inert gas such as He. The method involves positioning two carbon rods, end to end, with a separation of approximately 1mm. The carbon rods are housed in a glass enclosure, at low pressure and filled with inert gas. A direct current of 50-100 A is used to produce high temperature discharge between the two carbon rods. When pure graphite rods are used, fullerenes are deposited as soot inside the chamber, and multi-walled carbon nanotubes are deposited on the cathode. In the case of SWNT synthesis by an arc discharge method, the incorporation of catalytic metal particles in a **graphite anode** is necessary, and SWNTs are obtained as soot in an evaporation chamber.

Laser Ablation Synthesis:

Laser ablation means the **removal of material from a surface** by means of laser irradiation. The term "laser ablation" is used to emphasize the nonequilibrium vapor/plasma conditions created at the surface by intense laser pulse, to distinguish from "laser evaporation," which is heating and evaporation of material in condition of thermodynamic equilibrium. Employing a laser to ablate (gradually remove material from or erode the surface) graphite in an atmosphere containing an inert gas and a catalyst produces CNTs assembled to form ropes with diameters between 5 to 20nm diameter and tens to hundreds of micrometers long.

Chemical Vapor Deposition:

Chemical vapor deposition (CVD) is a chemical process used to produce high quality, high-performance, solid materials. The process is often used in the semiconductor industry to produce thin films. In typical CVD, the wafer (substrate) is exposed to one or more volatile precursors, which react and/or decompose on the substrate surface to produce the desired deposit. **Chemical vapor deposition (CVD) is** parent to a family of processes whereby a solid material is deposited from a vapor by a chemical reaction occurring on or in the vicinity of a normally heated substrate surface. The resulting solid material is in the form of a thin film, powder, or single crystal. **Chemical vapor deposition** definition is - a technique for **depositing** a usually thin solid layer of a substance on a surface as the result of **vapor-phase chemical** reactions in a high temperature gas near the surface. In short, hot vapors of any type of materials get deposited on the cold surface of another material.

Recommended videos:

https://www.bing.com/videos/search?q=Chemical+Vapor+Deposition+video&&v iew=detail&mid=5FE7E1727D6B3A4370B45FE7E1727D6B3A4370B4&&FOR M=VRDGAR

https://www.bing.com/videos/search?q=Chemical+Vapor+Deposition+video&&v iew=detail&mid=C2A55F29E1C04FADB2A8C2A55F29E1C04FADB2A8&rvs mid=5FE7E1727D6B3A4370B45FE7E1727D6B3A4370B4&FORM=VDRVRV

https://www.bing.com/videos/search?q=Chemical+Vapor+Deposition+video&&v iew=detail&mid=39791E9C51F2D520FAAE39791E9C51F2D520FAAE&&FO RM=VDRVRV

CNTs in Medicine:

The use of CNTs in drug delivery and biosensing technology has the potential to **revolutionize medicine**. Functionalization of single-walled nanotubes (SWNTs) has proven to enhance solubility and allow for efficient tumor targeting/drug delivery. It prevents SWNTs from being cytotoxic and altering the function of immune cells.

Medical implants made of porous plastic, coated with carbon nanotubes are being used for drug delivery. Therapeutic **drugs**, which are attached to the nanotubes can be released into the bloodstream, for example, when a change in the blood chemistry signals a problem. The **all-carbon** structure of **CNTs** allows for a wide variety of **chemical functionalization**, providing a platform that can be customized to a range of functions related to **regenerative medicine**.

Carbon nanotubes (CNTs) are very prevalent in today's world of medical research and are being highly explored in the fields of efficient drug delivery and biosensing methods for disease treatment and health monitoring. Carbon nanotube technology has shown to have the potential to alter drug delivery and biosensing methods for the better, and thus, carbon nanotubes have recently garnered interest in the field of medicine.

Cancer, a group of diseases in which cells grow and divide abnormally, is one of the primary diseases being looked at with regards to how it responds to CNT drug delivery. Current cancer therapy primarily involves surgery, radiation therapy, and chemotherapy. These methods of treatment are painful and kill normal cells in addition to producing adverse side effects. CNTs as drug delivery vehicles have shown potential in targeting specific cancer cells with a dosage lower than conventional drugs used, that is just as effective in killing the cells, however does not harm healthy cells and significantly reduces side effects. Current blood glucose monitoring methods by patients suffering from diabetes are normally invasive and often painful. For example, one method involves a continuous glucose sensor integrated into a small needle which must be inserted under the skin to monitor glucose levels every few days. Another method involves glucose monitoring strips to which blood must be applied. These methods are not only invasive, but they can also yield inaccurate results. It was shown that 70 percent of glucose readings obtained by continuous glucose sensors differed by 10 percent or more and 7 percent differed by over 50 percent. The high electrochemically accessible surface area, high electrical conductivity and useful structural properties have demonstrated the potential use of single-walled nanotubes (SWNTs) and multi-walled nanotubes (MWNTs) in highly sensitive noninvasive glucose detectors.

Graphene:

Graphene is an allotrope of carbon in the form of a two-dimensional, atomic-scale, hexagonal_lattice in which one atom forms each vertex. It is the basic structural element of other allotropes, including graphite, charcoal, carbon_nanotubes and fullerenes. It can also be considered as an indefinitely large aromatic molecule, the ultimate case of the family of flat polycyclic_aromatic hydrocarbons.

Graphene has many properties. In proportion to its thickness, it is about 100 times stronger than the strongest steel. It conducts heat and electricity very efficiently and is nearly transparent. Graphene also shows a large and nonlinear diamagnetism, even greater than graphite, and can be levitated by Nd-Fe-B magnets. Researchers have identified the bipolar transistor effect, ballistic_transport of charges and large quantum_oscillations in the material.

Graphene was discovered in 1947 with promises of extraordinary electric properties. But the problem was isolating a single sheet of material. For many years it was a material that only existed in theory and was believed to have an unstable nature. Then, the problem was solved, and graphene was born. The real material proved to be even better than was predicted. In truth, it was a miracle material.

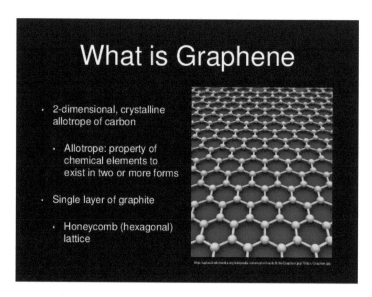

What is Graphene

- 2-dimensional, crystalline allotrope of carbon
  - Allotrope: property of chemical elements to exist in two or more forms
- Single layer of graphite
  - Honeycomb (hexagonal) lattice

Graphene is a sheet of crystalline carbon, with hexagonally arranged atoms. It is just one atom thick. It takes at least 10 layers before a sample becomes a bulk material. Graphene sheets are much stronger than steel, which is apparent when comparing tensile strength values. The difference is about three-four times.

Recommended videos:

https://www.bing.com/videos/search?q=Graphene+video&&view=detail&mid=4 B7A353A09F4E83C7D224B7A353A09F4E83C7D22&&FORM=VRDGAR

https://www.bing.com/videos/search?q=Graphene+video&&view=detail&mid=B E0B7FC2DA9B8A024FA8BE0B7FC2DA9B8A024FA8&rvsmid=4B7A353A09 F4E83C7D224B7A353A09F4E83C7D22&FORM=VDRVRV

https://www.bing.com/videos/search?q=Graphene+video&&view=detail&mid=7 BA0607E9FD4DF288E887BA0607E9FD4DF288E88&&FORM=VDRVRV

Graphene Synthesis:

Graphene has been **synthesized** in various ways and on different substrates. **Graphene** was first exfoliated mechanically from graphite in 2004. **Synthesis** of **graphene** refers to any process for fabricating or extracting **graphene**, depending on the desired size, purity and efflorescence of the specific product. In the earlier stage various **techniques** had been found for producing thin graphitic films.

Electrochemical **synthesis** can exfoliate **graphene**. Varying a pulsed voltage controls thickness, flake area, number of defects and affects its properties. The process begins by bathing the **graphite** in a solvent for intercalation. The process can be tracked by monitoring the solution's transparency with an LED and photodiode.

The other methods involve chemical reactions and graphite oxide. Another interesting method for preparation of graphene involves unzipping of CNTs. The most promising method is chemical vapor deposition onto metal substrates.

Recommended videos:

https://www.bing.com/videos/search?q=Graphene+Synthesis+video&&view=detail&mid=025407B1B95D0E8178D5025407B1B95D0E8178D5&&FORM=VDRVRV

Graphene Properties. Because of the strength of covalent bonds between carbon atoms, graphene has a **very high tensile strength**. (Basically, tensile relates to how much you can stretch something before it breaks.) In addition, graphene, unlike a buckyball or nanotube, has no inside because it is flat.

Electrical Property. One of the most useful properties of graphene is that it is a zero-overlap semimetal (with both holes and electrons as charge carriers) with very high electrical conductivity. Carbon atoms have a total of 6 electrons; 2 in the inner shell and 4 in the outer shell. The 4 outer shell electrons in an individual carbon atom are available for chemical bonding. In graphene, each atom is connected to 3 other carbon atoms on the two-dimensional plane, leaving 1 electron freely available in the third dimension for electronic conduction. These highly-mobile electrons are called **pi (π)** electrons and are located above and below the graphene sheet. These **pi** orbitals overlap and help to enhance the carbon to carbon bonds in graphene. Fundamentally, the electronic properties of graphene are dictated by the bonding and anti-bonding (the valance and conduction bands) of these **pi** orbitals.

Recommended videos:

https://www.bing.com/videos/search?q=Graphene+Properties+video&&view=detail&mid=FFCC2F3099D857B54206FFCC2F3099D857B54206&&FORM=VRDGAR

Mechanical Property. Another of graphene's stand-out properties is its inherent strength. Due to the strength of its 0.142 Nm - long carbon bonds, graphene is the strongest material ever discovered, with an ultimate tensile strength of 130,000,000,000 Pascals (or 130 gigapascals), compared to 400,000,000 for A36 structural steel, or 375,700,000 for Aramid (Kevlar). Not only is graphene extraordinarily strong, it is also very light at 0.77 milligrams per square meter (for comparison purposes, 1 square meter of paper is roughly 1000 times heavier). It is often said that a single sheet of graphene (being only 1 atom thick), sufficient in size enough to cover a whole football field, would weigh under 1 single gram.

What makes this particularly special is that graphene also contains elastic properties, being able to retain its initial size after strain.

Recommended videos:

https://www.bing.com/videos/search?q=Graphene+Properties+video&&view=detail&mid=7E9D4CA3B8B6860D05D77E9D4CA3B8B6860D05D7&&FORM=VRDGAR

Optical Properties. Graphene's ability to absorb a rather large 2.3% of white light is also a unique and interesting property, especially considering that it is only 1 atom thick. This is due to its aforementioned electronic properties; the electrons acting like massless charge carriers with very high mobility. A few years ago, it was proved that the amount of white light absorbed is based on the Fine Structure Constant, rather than being dictated by material specifics. Adding another layer of graphene increases the amount of white light absorbed by approximately the same value (2.3%).

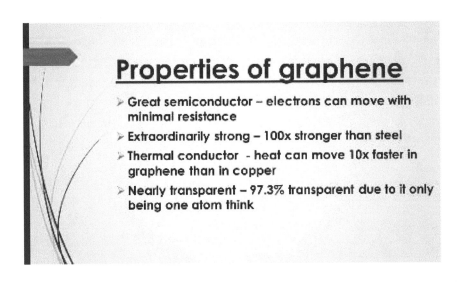

Recommended videos:

https://www.bing.com/videos/search?q=Graphene+Properties+video&&view=detail&mid=48444A62B1FF2653980448444A62B1FF26539804&&FORM=VDRVRV

https://www.bing.com/videos/search?q=Graphene+Properties+video&&view=detail&mid=5A1220E2333C9BCD46B35A1220E2333C9BCD46B3&&FORM=VDRVRV

5.0   Molecule-Based Nanotechnology

Introduction:

The natural world has been employing molecular materials for the assembly of structures for billions of years. **Molecular nanotechnology** (MNT) is a technology **based** on the ability to build structures to complex, atomic specifications by means of mechanosynthesis. This is distinct from nanoscale materials.

The vision of **revolutionary bottom-up nanotechnology** is **based** on a concept of **molecular assembly technologies** where **nanoscale** materials and structures self-assemble to **microscale** structures and finally to macroscopic devices and products.

Recommended videos:

https://www.bing.com/videos/search?q=Molecule-Based+Nanotechnology+video&&view=detail&mid=82C18A18B3170E9D8BC182C18A18B3170E9D8BC1&&FORM=VRDGAR

https://www.bing.com/videos/search?q=Molecule-Based+Nanotechnology+video&&view=detail&mid=D411E3F5D329AE2EE49AD411E3F5D329AE2EE49A&&FORM=VRDGAR

DNA Nanotechnology:

DNA is a beautifully clever and complex molecule that holds all the genetic information necessary for growth and development, functioning and reproduction of all known living things. But DNA nanotechnology is not interested in these genetic instructions – instead it focusses on the design, study and application of synthetic structures that are based on DNA, making use of the nucleic acid's physical and chemical characteristics.

DNA nanotechnology is the design and manufacture of artificial nucleic acid structures for technological uses. In this field, nucleic acids are employed as non-biological engineering materials for nanotechnology rather than as the carriers of genetic information in living cells.

Structural **DNA Nanotechnology** uses unusual **DNA** motifs to build target shapes and arrangements. These unusual motifs are generated by reciprocal exchange of **DNA backbones**, leading to branched systems with many strands and multiple helical domains. The motifs may be combined by sticky ended cohesion, involving hydrogen bonding or covalent interactions.

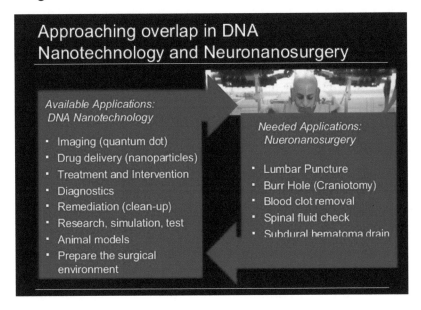

Recommended videos:

https://www.bing.com/videos/search?q=DNA+Nanotechnology+video&&view=detail&mid=2E00E1892CBADB8A25662E00E1892CBADB8A2566&&FORM=VRDGAR

https://www.bing.com/videos/search?q=DNA+Nanotechnology+video&&view=detail&mid=1A103331E5A44390FD2F1A103331E5A44390FD2F&&FORM=VDRVRV

https://www.bing.com/videos/search?q=DNA+Nanotechnology+video&&view=detail&mid=6F3BAF2C87D69BE0E1CD6F3BAF2C87D69BE0E1CD&&FORM=VRDGAR

Self-Assembled Monolayers:

Self-assembled monolayers (SAMs) refer to a single layer of molecules that assemble themselves in an orderly manner on a solid surface. Traditional formation of **SAMs** may be modified to manipulate NP surfaces for various applications. Self-assembled monolayers (SAM) of organic molecules are molecular assemblies formed spontaneously on surfaces by adsorption and are organized into more or less large ordered domains. In some cases, molecules that form the monolayer do not interact strongly with the substrate. This is the case for instance of the two-dimensional supramolecular networks of e.g.

Self-assembled monolayers are now established as crucial **interlayers** and electronically active layers in organic electronic devices, such as organic light emitting diodes (OLEDs), **organic photovoltaics (OPVs)**, organic thin film transistors (OTFTs), and **nonvolatile memories** (**NVMs**). SAM on gold are the most thoroughly studied monolayer system because gold is inert and alkanethiol molecules produce well-ordered SAMs on gold. **Alkanethiol** definition is - any of various hydrocarbon polymers also containing sulfur that are deposited on a surface (such as gold) in a monolayer typically in order to create an electronic circuit. Additionally, gold does not form an oxide at room temperature. SAMs are formed by immersing substrates, such as gold, in dilute solutions of alkanethiols.

The rate of each step depends on experimental factors such as the concentration of the solution and the length of the carbon chain. It is important that the length of the alkanethiol carbon chain also affects how well the SAM is organized. With longer chain alkanethiol molecules and chains standing up, the number of defective sides in the final molecule film is reduced. Defects appear in the resultant SAM when alkanethiol molecules with smaller carbon chains are used due to weaker London dispersion forces acting between neighboring molecules.

Alkanethiol solution   Alkanethiol molecules   Organization of
                              attaching to gold   alkanethiols on gold

SAMs are created by the chemisorption of "head groups" onto a substrate from either the vapor or liquid phase followed by a slow organization of "tail groups". Initially, at small molecular density on the surface, adsorbate molecules form either a disordered mass of molecules or form an ordered two-dimensional "lying down phase", and at higher molecular coverage, over a period of minutes to hours, begin to form three-dimensional crystalline or semi-crystalline structures on the substrate surface.

SAM add functionality to nanometer-scale objects such as nanoparticles and nanowires. SAM functionality allows researchers to engineer surface properties and localize chemical reactions. For example, SAM can make a surface receptive to specific types of molecules or render surfaces chemically inert. It is noted that SAMs may provide pathway to nanoscale devices to be used in "organic electronics".

"Organic electronics" is a field of materials science concerning the design, synthesis, characterization, and application of organic small molecules or polymers that show desirable electronic properties such as conductivity. Unlike conventional inorganic conductors and semiconductors, organic electronic materials are constructed from organic (carbon-based) small molecules or polymers using synthetic strategies developed in the context of organic and polymer chemistry. One of the promised benefits of organic electronics is their potential low cost compared to traditional inorganic electronics. Attractive properties of polymeric conductors include their electrical conductivity that can be varied by the concentrations of dopants. Relative to metals, they have mechanical flexibility. Some have high thermal stability.

Dendrimers:

Dendrimers a synthetic polymer with a structure of repeatedly branching chains, typically forming spherical macromolecules. **Dendrimers** are repetitively branched molecules. The name comes from the Greek word (dendron), which translates to "tree". Synonymous terms for dendrimer include arborols and cascade molecules. However, dendrimer is currently the internationally accepted term. A dendrimer is typically symmetric around the core, and often adopts a spherical three-dimensional morphology.

Dendrimers are highly branched, star-shaped macromolecules with nanometer-scale dimensions. Dendrimers are defined by three components: a central core, an interior dendritic structure (the branches), and an exterior surface with functional surface groups. The varied combination of these components yields products of different shapes and sizes with shielded interior cores that are ideal candidates for applications in both biological and materials sciences. While the attached surface groups affect the solubility and chelation ability, the varied cores impart unique properties to the cavity size, absorption capacity, and capture-release characteristics.

The structure of dendrimers has a significant impact on their chemical and physical properties. Dendrimers are usually about 5nm across. Due to the extensive network of branches in dendrimer molecules, many applications are possible including, catalysis, electronics, and drug release.

Recommended videos:

https://www.bing.com/videos/search?q=Dendrimers+video&&view=detail&mid=E248E5C7C043FC62815FE248E5C7C043FC62815F&&FORM=VDRVRV

https://www.bing.com/videos/search?q=dendrimers+nanotechnology&&view=detail&mid=F09E907D45DCCDDCE905F09E907D45DCCDDCE905&&FORM=VDRVRV

Applications of dendrimers:

Applications highlighted in recent literature include drug delivery and control, diagnostics, gene transfection, catalysis, energy harvesting, photo activity, molecular weight and size determination, rheology modification, and nanoscale science and technology, and much more. New ideas come almost on regular basis.

Dendrimer synthesis:

Monodisperse dendrimers are synthesized by step-wise chemical methods to give distinct generations (G0, G1, G2, ...) of molecules with narrow molecular weight distribution, uniform size and shape, and multiple (multivalent) surface groups Z. Dendrons are monodisperse wedge-shaped sections of dendrimers with a single focal point reactive function. Hyperbranched polymers are polydisperse dendritic macromolecules synthesized by lower-cost polymerization methods.

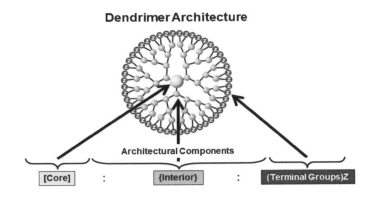

Dendrimers contain three different regions: the core, branches, and the terminal group. The core is a single atom or molecule dendrons are attached to. The dendrons (branches) are monomers linked to the core, forming layers and building successive generations. Branched monomers, dendrons, extend from the core. At the end of the branches are terminal groups, which can be easily modified to change the molecule's chemical and physical properties such as chemical reactivity, stability, solubility, and toxicity.

By selecting proper terminal groups dendrimers can be tailored for specific uses. For instance, the solubility of dendrimers is strongly influenced by the nature of the terminal groups. Hydrophilic functional groups increase the solubility of dendrimers in polar solvents, while dendrimers with hydrophobic functional groups are soluble in nonpolar solvents.

A **hydrophile** is a molecule or other molecular entity that is attracted to and tends to be dissolved by water. A hydrophilic molecule or portion of a molecule is one whose interactions with water and other polar substances are more thermodynamically favorable than their interactions with oil or other **hydrophobic** solvents. They are typically charge-polarized and capable of hydrogen bonding. Hydrophilic functional groups include **hydroxyl groups** (resulting in alcohols though also found in sugars, etc.), carbonyl groups (giving rise to aldehydes and ketones), carboxyl groups (resulting in carboxylic acids), amino groups (i.e., as found in amino acids), sulfhydryl groups (giving rise to thiols, i.e., as found in the amino acid cysteine), phosphate groups (as found in nucleic acids and phospholipids), etc.

Recommended videos:

https://www.bing.com/videos/search?q=Dendrimer+Drug+Delivery&&view=detail&mid=9D1EBE6360FB7576C6C19D1EBE6360FB7576C6C1&&FORM=VRDGAR

https://www.bing.com/videos/search?q=Dendrimer+Drug+Delivery&&view=detail&mid=72C769473A9E78FD17AB72C769473A9E78FD17AB&rvsmid=9D1EBE6360FB7576C6C19D1EBE6360FB7576C6C1&FORM=VDRVRV

Divergent and Convergent Dendrimer Synthesis:

The **synthesis** of **dendrimers** follows either a **divergent** or **convergent** approach. **Dendrimers** can be **synthesized** by two major approaches. In the **divergent** approach, used in early periods, the **synthesis** starts from the core of the **dendrimer** to which the arms are attached by adding building blocks in an exhaustive and step-wise manner. In the **convergent** approach, first suitable building blocks are prepared and then connected by Lego chemistry, ligations or click chemistry. In the **divergent** approach, the **synthesis** is done stepwise.

Convergent and **divergent dendrimer synthesis** strategies basically represent the two approaches to stepwise growth. However, it has been pointed out that these two strategies can effectively be combined into one set of reactions termed "double exponential **dendrimer** growth".

There are two defined methods of **dendrimer synthesis, divergent synthesis** and **convergent synthesis**. However, because the actual reactions consist of many steps needed to protect the active site, it is difficult to synthesize dendrimers using either method. This makes dendrimers hard to make and very expensive to purchase.

Recommended videos:

https://www.bing.com/videos/search?q=Dendrimer+synthesis&&view=detail&mid=D5AFF18900E71B004A0FD5AFF18900E71B004A0F&&FORM=VRDGAR

https://www.bing.com/videos/search?q=Divergent+and+Convergent+Dendrimer+Synthesis&&view=detail&mid=3CC2C5F78181F1F0F1843CC2C5F78181F1F0F184&&FORM=VRDGAR

Dendrimers in Medicine:

Dendrimers are three-dimensional macromolecular structures originating from a central core molecule and surrounded by successive addition of branching layers (generation). These structures exhibit a high degree of molecular uniformity, narrow molecular weight distribution, tunable size and shape characteristics, as well as multivalence. Collectively, these physicochemical characteristics together with advancements in design of biodegradable backbones have conferred many applications to dendrimers in formulation science and nano-pharmaceutical developments. These have included the employment of dendrimers as pro-drugs and vehicles for solubilization, encapsulation, complexation, delivery, and site-specific targeting of small-molecule drugs, biopharmaceuticals, and contrast agents.

Some Examples of Products that Use Dendrimers:

Dendrimer-based products (and those in the pipeline) include, for example:

- A dendrimer-based tool for detecting cardiac damage is being developed by Dade Behring, one of the world's largest medical diagnostic firms;
- The world's first drug based on dendrimers, developed by Australian-based StarPharma, is a topical gel for use as a "liquid condom" to reduce the risk of HIV infection in women. StarPharma's "VivaGel" microbicide has gone through initial animal testing and phase-one safety trials in humans;

- The US Army Research Laboratory is developing a dendrimer-based anthrax detection agent, dubbed "Alert Ticket";
- ExxonMobil owns patent 5,906,970 on a "flow improver" based on dendrimer technology - an additive that will increase the flow of oil in cold temperatures.

Recommended videos:

https://www.bing.com/videos/search?q=Applications+of+Dendrimers&&view=detail&mid=87EC9C3FF8034BA47D6F87EC9C3FF8034BA47D6F&&FORM=VRDGAR

https://www.bing.com/videos/search?q=Applications+of+Dendrimers&&view=detail&mid=28D3891AE62AEB06131128D3891AE62AEB061311&&FORM=VDRVRV

https://www.bing.com/videos/search?q=Dendrimer+Drug+Delivery&&view=detail&mid=39319E7712598C6229E039319E7712598C6229E0&&FORM=VDRVRV

Lipids:

Lipids are any of a class of organic compounds that are fatty acids or their derivatives and are insoluble in water but soluble in organic solvents. They include many natural oils, waxes, and steroids. Lipids, also known as **fats**, are one of the four macromolecules. They are an organic compound, which means they contain carbon atoms. There are many types of lipids with various functions. One function of lipids is for long term energy.

Lipids are **fats**. In the body they take the form of phospholipids, cholesterol and fatty acids. Although fats play a role in obesity and disease, your body needs a certain amount of fat to function - also known as **essential body fat.**

Lipids have a range of desirable properties for use in nanotechnology. They can self-assemble into nanofilms and other nanostructures such as micelles, reverse micelles, and liposomes. Additionally, lipid assemblies can be attached to other nanostructures via specific chemical linkages. These features along with transparency of lipid structures in visible light and their heat conductivity, have made lipids an important building block for nanotechnology. Lipid molecules are 2-4nm in size.

Recommended videos:

https://www.bing.com/videos/search?q=Lipids+videos&&view=detail&mid=71C06820DFA739CE097A71C06820DFA739CE097A&&FORM=VRDGAR

https://www.bing.com/videos/search?q=Lipids+videos&&view=detail&mid=11B38647D0EA5AC425A211B38647D0EA5AC425A2&&FORM=VRDGAR

https://www.bing.com/videos/search?q=Lipids+videos&&view=detail&mid=BD45BD0DE79B1C4FCC13BD45BD0DE79B1C4FCC13&&FORM=VDRVRV

Micelles:

A micelle or micella (plural micelles or micellae, respectively) is an aggregate of surfactant molecules dispersed in a liquid colloid. A typical micelle in aqueous solution forms an aggregate with the hydrophilic "head" regions in contact with surrounding solvent, sequestering the hydrophobic single-tail regions in the micelle center.

Colloid a homogeneous non-crystalline substance consisting of large molecules or ultramicroscopic particles of one substance dispersed through a second substance. Colloids include gels, sols, and emulsions; the particles do not settle, and cannot be separated out by ordinary filtering or centrifuging like those in a suspension.

Micelles are structures formed from the self-assembly of molecules known as surfactants. Surfactants are molecules with hydrophobic tail and hydrophilic head groups. Thus, two sides of these molecules would react differently but predictably. Amphiphilic molecules can form **micelles** not only in water, but also in nonpolar organic solvents. In such cases, micelle aggregates are called inverse **micelles** because the situation is inverted as respect to water. In fact, hydrocarbon tails are exposed to the solvent, while the polar heads point toward the interior of the aggregate to escape the contacts with the solvent.

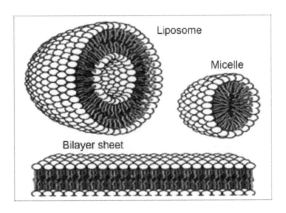

Recommended videos:

https://www.bing.com/videos/search?q=Micelles+videos&&view=detail&mid=E3
A9B4BE80BB708B9B50E3A9B4BE80BB708B9B50&&FORM=VRDGAR

https://www.bing.com/videos/search?q=Micelles+videos&&view=detail&mid=3C
634A93B72E98D7DB763C634A93B72E98D7DB76&rvsmid=E3A9B4BE80BB7
08B9B50E3A9B4BE80BB708B9B50&FORM=VDRVRV

https://www.bing.com/videos/search?q=Micelles+videos&&view=detail&mid=F2
272C8D9901827B028BF2272C8D9901827B028B&&FORM=VRDGAR

Molecular Electronics:

Molecular electronics is a branch of electronics in which individual molecules perform the same function as microelectronic devices such as diodes. Molecular electronics is the study and application of molecular building blocks for the fabrication of electronic components. It is an interdisciplinary area that spans physics, chemistry, and materials science. The unifying feature is the use of molecular building blocks for the fabrication of electronic components.

Essentially all electronic processes in nature, from photosynthesis to signal transduction, occur in molecular structure. Molecular electronics can be defined as technology utilizing single molecules, small groups of molecules, or carbon nanotubes to perform electronic functions.

Do molecular compounds conduct electricity? A molecular compound **cannot conduct electricity in any state**, whereas an ionic compound, when dissolved in an aqueous solution, can act as a good conductor of electricity. Ionic compounds are more reactive than molecular compounds.

What are molecular motors? Molecular motors are **biological molecular machines that are the essential agents of movement in living organisms**. In general terms, a motor is a device that consumes energy in one form and converts it into motion or mechanical work; for example, many protein -based molecular motors harness the chemical free energy released converting it into movements or senses.

What is a molecular computer? Molecular computing is a generic term for any **computational scheme** which uses individual atoms or molecules as a means of solving computational problems. These computers, when ready for the consumer use would be thousands of times faster and contain more power. Quantum and molecular computers are the computers of not so distant future.

Although **molecular electronics** has been the subject of research for some time, over the past few years a number of synthetic and quantum chemists, physicists, engineers, and other researchers have sharply increased the ranks of this field. Several new **molecular-**

electronic systems, analytical tools, and device architectures have been developed and employed in research and the industry.

Molecular-based electronics are advantageous because of their low cost, light weight, mechanically flexibility, and excellent compatibility with plastic substrates. Advanced organic electron systems are already utilized in commercial products with efficiency, bright, and colorful thin display.

Recommended videos:

https://www.bing.com/videos/search?q=Molecular+Electronics+videos&&view=detail&mid=55E7DD68406D4C27BB1955E7DD68406D4C27BB19&&FORM=VRDGAR

https://www.bing.com/videos/search?q=Molecular+Electronics+videos&&view=detail&mid=83BE59EAA8682B1D5B1F83BE59EAA8682B1D5B1F&&FORM=VDRVRV

https://www.bing.com/videos/search?q=Molecular+Electronics+videos&&view=detail&mid=94240C961DBC09A3BBFC94240C961DBC09A3BBFC&&FORM=VRDGAR

OLEDs:

OLED is a light-emitting diode containing thin flexible sheets of an organic electroluminescent material, used mainly in digital display screens. **OLEDs** are used to create digital displays in devices such as television screens, computer monitors, portable systems such as cell and smartphones, handheld game consoles and PDAs. A major area of research is the development of white OLED devices for use in solid-state lighting applications. Like any display, **OLEDs** have limited lifetime, that was quite a problem a few years ago. But there has been constant progress, and today this is almost a non-issue. Today **OLEDs** last long enough to be used in mobile devices and TVs. **OLEDs** can also be problematic in direct sunlight, because of their emissive properties.

OLEDs are solid-state devices composed of thin films of organic molecules that create light with the application of electricity. **OLEDs** can provide brighter, crisper displays on electronic devices and use less power than conventional light-emitting diodes (LEDs) or liquid crystal displays (LCDs) used today.

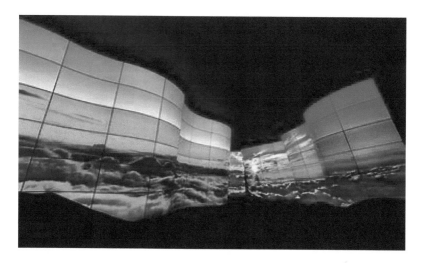

Another attractive feature of organic electronics is the ability to deposit organic films on low-cost substrates such as glass, plastic or metal foil. This is due to the relatively ease of processing of the organic compounds. These organic compounds can be tailored to optimize a function, such as charge mobility or luminescent properties.

Recommended videos:

https://www.bing.com/videos/search?q=OLEDs+videos&&view=detail&mid=E9 7489ECD9C0CAB13084E97489ECD9C0CAB13084&&FORM=VRDGAR

https://www.bing.com/videos/search?q=OLEDs+videos&&view=detail&mid=3F A0E4A91D1A015A3A123FA0E4A91D1A015A3A12&&FORM=VDRVRV

https://www.bing.com/videos/search?q=OLEDs+videos&&view=detail&mid=691 C491071D603BDE1DF691C491071D603BDE1DF&&FORM=VDRVRV

https://www.bing.com/videos/search?q=OLEDs+videos&&view=detail&mid=0F A15A17E4BB0E6F16730FA15A17E4BB0E6F1673&&FORM=VRDGAR

Organic Solar Cells:

An organic solar cell or plastic solar cell is a type of photovoltaic that uses organic electronics, a branch of electronics that deals with conductive organic polymers or small organic molecules, for light absorption and charge transport to produce electricity from sunlight by the photovoltaic effect. An example of an organic photovoltaic is the polymer solar cell.

The amount of energy that the Earth receives from the sun is enormous: 1.75×1017 W. As the world energy consumption in 2003 amounted to 4.4×1020 J, Earth receives enough energy to fulfill the yearly world demand of energy in less than an hour. Not all of that energy reaches the Earth's surface due to absorption and scattering, and the photovoltaic conversion of solar energy remains an important challenge. State-of-the-art inorganic solar cells have a record power conversion efficiency of close to 39%, while commercially available solar panels, have a significantly lower efficiency of around 15–20%. Another approach to making solar cells is to use organic materials, such as conjugated polymers.

Solar cells based on thin polymer films are particularly attractive because of their ease of processing, mechanical flexibility, and potential for low cost fabrication of large areas. Additionally, their material properties can be tailored by modifying their chemical makeup, resulting in greater customization than traditional solar cells allow. Although significant progress has been made, the efficiency of converting solar energy into electrical power obtained with plastic solar cells still does not warrant commercialization: the most efficient devices have an efficiency of 4-5%. To improve the efficiency of plastic solar cells it is crucial to understand what limits their performance.

Organic solar cells can be distinguished by the production technique, the character of the materials and by the device design. The two main production techniques can be identified as either wet processing or thermal evaporation. Device architectures are single layer, bi-layer hetero junction and bulk hetero junction, with the diffuse bi-layer hetero junction as intermediate between the bi-layer and the bulk hetero junction. Whereas the single layer comprises of only one active material, the other architectures are based on respectively two kinds of materials: electron donors (D) and electron acceptors (A).

Organic materials bear the potential to develop a long-term technology that is economically viable for large-scale power generation based on environmentally safe materials with unlimited availability. Organic semiconductors are a less expensive alternative to inorganic semiconductors like Si. They can have extremely high optical absorption coefficients which offer the possibility to produce ultra-thin solar cells. Additional attractive features of organic PVs are the possibilities for thin flexible devices which can be fabricated using high throughput, low temperature approaches that employ well established printing techniques in a roll-to-roll process This possibility of using flexible plastic substrates in an easily scalable high-speed printing process can reduce the balance of system cost for organic PVs, resulting in a shorter energetic pay-back time. The electronic structure of all organic semiconductors is based on conjugated π-electrons. A conjugated organic system is made of an alternation between single and double carbon-carbon bonds. Single bonds are known as σ-bonds and are associated with localized electrons, and double bonds contain a σ-bond and a π-bond. The π-electrons are much more mobile than the σ-electrons; they can jump from site to site between carbon atoms thanks to the mutual overlap of π orbitals along the conjugation path, which causes the wave functions to delocalize over the conjugated backbone. The π-bands are either empty (called the Lowest Unoccupied Molecular Orbital - LUMO) or filled with electrons (called the Highest Occupied Molecular Orbital - HOMO). The band gap of these materials ranges from 1 to 4 eV. This π-electron system has all the essential electronic features of organic materials: light absorption and emission, charge generation and transport.

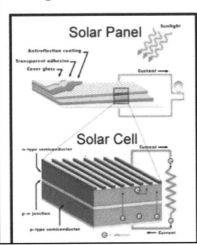

Organic Photovoltaic Cells

- Organic photovoltaic cells are solar cells that utilize organic polymers and small molecules as the active layer for light absorption and charge transport

Organic solar cell research has developed during the past 30 years, but especially in the last decade it has attracted scientific and economic interest triggered by a rapid increase in power conversion efficiencies. This was achieved by the introduction of new materials, improved materials engineering, and more sophisticated device structures.

Recommended videos:

https://www.bing.com/videos/search?q=Organic+Solar+Cells+videos&&view=detail&mid=72B1B56F4C6670F0FC5072B1B56F4C6670F0FC50&&FORM=VRDGAR

https://www.bing.com/videos/search?q=Organic+Solar+Cells+videos&&view=detail&mid=40BF4118503D8058983D40BF4118503D8058983D&&FORM=VDRVRV

https://www.bing.com/videos/search?q=Organic+Solar+Cells+videos&&view=detail&mid=17894E655C1CBEBE0FE617894E655C1CBEBE0FE6&&FORM=VRDGAR

Dye-Sensitized Solar Cells (DSSCs):

What is a Dye-Sensitized Solar Cell? **Dye-sensitized solar cells (DSSC)** are an efficient type of **thin-film photovoltaic cell**. They are based on printable semiconductor materials, including an electrolyte and a **photo-sensitized anode**.

A dye-sensitized solar cell (DSSC, DSC, DYSC or Grätzel cell) is a **low-cost solar cell belonging to the group of thin film solar cells**. It is based on a semiconductor formed between a photo-sensitized anode and an electrolyte, a photoelectrochemical system.

Dye Sensitized solar cells (DSSC), also sometimes referred to as **dye sensitized cells** (DSC), are a **third generation photovoltaic** (solar) **cell** that converts any visible light into electrical energy. This new class of advanced **solar cell** can be likened to artificial photosynthesis due to the way in which it mimics natural absorption of light energy. The mechanism of dye-sensitized **solar cells** can be understood through **equivalent circuits**, which are considered to be useful tools to analyze **cell** devices and improve **cell** performance. The information obtained from the equivalent circuit model of **DSSCs** includes electrical processes, system simulation, and **cell** modules.

In electrical engineering and science, an **equivalent circuit** refers to a theoretical circuit that retains all of the electrical characteristics of a given circuit. Often, an equivalent circuit is sought that simplifies calculation, and more broadly, that is a simplest form of a more complex circuit in order to aid analysis.

Dye sensitized solar cell (DSSC) is the only **solar cell** that can offer both the flexibility and transparency. Its efficiency is comparable to **amorphous silicon solar cells** but with a much lower cost. This review not only covers the fundamentals of **DSSC** but also the related cutting-edge research and its development for industrial applications.

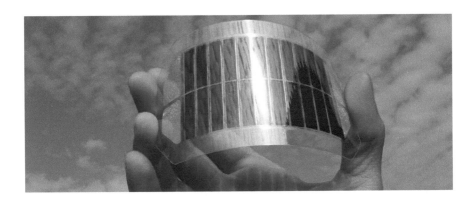

As a novel **photovoltaic technology, dye-sensitized solar cells** (**DSSCs**) have potential to compete with traditional **solar cells**. Materials such as **TiO 2** used in **DSSCs** are generally inexpensive, abundant and innocuous to the environment. Compared with **silicon solar cells**, they are insensitive to impurities in fabrication process, which accelerates a transition from research laboratory to mass production line.

Recommended videos:

https://www.bing.com/videos/search?q=Dye-Sensitized+Solar+Cells+(DSSCs)+videos&&view=detail&mid=A92F22EACBEB558406D0A92F22E ACBEB558406D0&&FORM=VRDGAR

https://www.bing.com/videos/search?q=Dye-Sensitized+Solar+Cells+(DSSCs)+videos&&view=detail&mid=EDB0CB3BB0A971D866BBEDB0C B3BB0A971D866BB&&FORM=VRDGAR

https://www.bing.com/videos/search?q=Dye-Sensitized+Solar+Cells+(DSSCs)+videos&&view=detail&mid=5D1ADA2C49A525D66CA15D1ADA 2C49A525D66CA1&&FORM=VRDGAR

Drug Delivery Systems:

Certain nanoscale materials have allowed biological tests to be performed quickly, with improved sensitivity and increased flexibility. With the help of nanotechnology early detection, prevention, improved diagnosis, and proper treatment is possible. Recent developments in nanotechnology have shown nanomaterials to have a great potential as drug carriers. For this reason, a variety of nanoparticles such as dendrimers and nano-porous materials find applications in the medical field.

Development of an effective approach for delivering a new drug is as important as inventing a new drug. Even if a developed new drug has excellent pharmacokinetic and therapeutic properties, it shows its potential activity in the body effectively only when it is exactly targeted to specific molecules. Various nanotechnological approaches for effective drug delivery have been developed and some of them have already been successfully commercialized. Most prominent nano-drug delivery systems that are in market place are oncology related and based on liposomal, solid nanoparticle based, protein polymer conjugates and polymer-drug conjugate-based delivery platforms.

The bioavailability (problem of the conventional drug delivery) of a drug within the body depends on several factors like the size of the drug molecules and solubility parameters. Conventional dosage forms therefore face challenges in reaching the target site at appropriate dose. For example, conventional dosage forms of some of the highly water-soluble drugs cause fluctuations in drug concentration in the body due to high disintegration properties and result in faster clearance of the drug from the blood stream. Other drugs are fat soluble and when taken in conventional dosage forms may cause bioavailability problems. Similarly, patients suffering from chronic diseases like diabetes need to take painful insulin injections on a regular basis. Also, cancer patients regularly must undergo powerful chemotherapy, which involves quite severe side effects as the anticancer drugs target cancer cells and normal cells equally.

Hence proper platforms to deliver the drugs at targeted sites without losing their efficacies while limiting the associated side effects are highly required. Many novel technologies for developing effective drug delivery systems came into existence among nanotechnology platforms for achieving targeted drug delivery. Research in this field includes the development of drug nanoparticles, polymeric and inorganic biodegradable nanocarriers for drug delivery, and surface engineering of carrier molecules. These nanocarriers help in solubilizing the lipophilic drugs, protecting fragile drugs from enzymatic degradation, pH conditions, etc., and targeting specific sites with triggered release of drug contents.

Recommended videos:

https://www.bing.com/videos/search?q=nano+Drug+Delivery+Systems+videos&&view=detail&mid=D0F611852A6D2394F0EED0F611852A6D2394F0EE&&FORM=VRDGAR

https://www.bing.com/videos/search?q=nano+Drug+Delivery+Systems+videos&&view=detail&mid=9508C1C51E7A028F7C069508C1C51E7A028F7C06&&FORM=VDRVRV

Nanobots:

Nanobots or nanomotors are advanced sub-micron sized, self-driven, biodegradable nanodevices made of bio-nano components, which carry cargo to the target sites. This active motor-based drug delivery approach promises an effective and improved drug delivery compared to conventional methods. Gold nanoparticle loaded PEDOT/zinc-based artificial micromotors are tested in mouse models via oral administration.

They showed excellent acid-driven, self-propulsive properties with high cargo-loading capacities. Unimolecular submersible nanomachines that are activated by UV light, DNA-origami based nanorobots, light-induced actuating nano-transducers, WiNoBots, magnetic multilink nano-swimmers, etc., are some of the other technological developments that are anticipating the application of nanorobots in drug delivery.

Recommended videos:

https://www.bing.com/videos/search?q=Nanobots+videos&&view=detail&mid=4E973E5DA32B1C108A074E973E5DA32B1C108A07&&FORM=VRDGAR

https://www.bing.com/videos/search?q=Nanobots+videos&&view=detail&mid=DF21E6CB2D5999F5DDEDDF21E6CB2D5999F5DDED&&FORM=VDRVRV

Nanoghosts:

Nanoghost technology is one of the latest approaches developed for smart drug/gene delivery. Nanoghosts are a type of nanovesicles derived from naturally functionalized mammalian cell surface membranes of whole biological cells. These naturally derived carriers overcome drug loading issues, evade tumor specific immune responses, provide greater nanoparticle stability and improve drug release profiles.

Recommended videos:

https://www.bing.com/videos/search?q=Nano+ghosts+drug+delivery&&view=detail&mid=8516F31FD0BF8BA174A88516F31FD0BF8BA174A8&&FORM=VRDGAR

https://www.bing.com/videos/search?q=Nano+ghosts+drug+delivery&&view=detail&mid=53AD8E0507B66B62AD6B53AD8E0507B66B62AD6B&&FORM=VDRVRV

Nanoclews:

Nanoclew or nanococoon is a DNA based biocompatible drug delivery system. In this system single stranded DNA makes whole nanococoon. It self assembles to look like a yarn or cocoon or a clew like structure by rolling-circle amplification. Still, it is only single stranded DNA. Biomedical engineering researchers also developed a drug delivery system consisting of nanococcons made of DNA that target cancer cells and trick the cells into absorbing the cocoon before unleashing anticancer drugs.

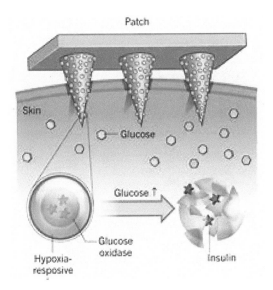

Recommended videos:

https://www.bing.com/videos/search?q=Nano+ball+drug+delivery&&view=detail&mid=B03114917614713D1C9CB03114917614713D1C9C&&FORM=VRDGAR

https://www.bing.com/videos/search?q=Nano+ball+drug+delivery&&view=detail&mid=B7DC80FF2272DB3D3C8FB7DC80FF2272DB3D3C8F&&FORM=VDRVRV

Nanoneedles:

Direct delivery of therapeutic molecules to cell cytoplasm is highly desirable in drug delivery for achieving high efficacy and reduced side effects. But biological membranes act as strong barriers for entry of drug into the cell. Nanoneedles could help to overcome these problems. Nanoneedles are very small and make temporary perforations to biological membranes. Hence these needles, without disturbing biological functions of the body can deliver the drugs. These nanoneedles are needles and drug carriers at the same time and that is most efficient and the least expensive way to talk the problem.

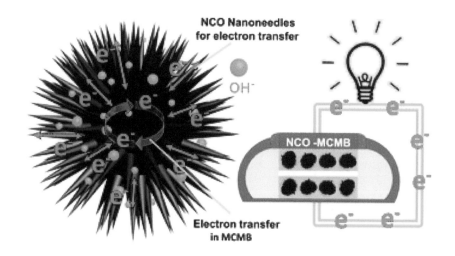

Recommended videos:

https://www.bing.com/videos/search?q=Nanoneedles&&view=detail&mid=D009731E366535A5DF4DD009731E366535A5DF4D&&FORM=VRDGAR

https://www.bing.com/videos/search?q=Nanoneedles&&view=detail&mid=ED7A0D8109E97BF5228EED7A0D8109E97BF5228E&&FORM=VDRVRV

Nanoclusters:

Metal nanoclusters are self-assembled nanoparticles made of polymers or small organic molecules crosslinked with plasmonic metals (such as gold, silver, or magnetic particles). Because of their molecular-like and fluorescence properties, they have gained importance in the field of drug delivery as well as biosensing and bioimaging.

Recommended videos:

https://www.bing.com/videos/search?q=Nanoclusters&&view=detail&mid=A6D3FDC19615D2257B9DA6D3FDC19615D2257B9D&&FORM=VDRVRV

https://www.bing.com/videos/search?q=Nanoclusters&&view=detail&mid=D34676FA355A0FADE4A4D34676FA355A0FADE4A4&&FORM=VDRVRV

https://www.bing.com/videos/search?q=Nanoclusters&&view=detail&mid=D1EA434571B9609D8504D1EA434571B9609D8504&&FORM=VDRVRV

Nanobubbles:

Nanobubbles are gas-filled spherical nano-sized structures often stabilized by polymeric/lipid shells. These nanocarriers in combination with thermal, ultrasound, acoustic or magnetic sensitivities are employed as more efficient imaging and drug delivery agents in various therapeutic treatments. They are more stable and showed longer residence time in systemic circulation.

Recommended videos:

https://www.bing.com/videos/search?q=Nanobubbles&&view=detail&mid=8570 28B768BBDF51AA91857028B768BBDF51AA91&&FORM=VRDGAR

https://www.bing.com/videos/search?q=Nanobubbles&&view=detail&mid=FF4 21DAE67CF7B51AF39FF421DAE67CF7B51AF39&&FORM=VDRVRV

Exosomes:

Exosomes - tiny biological nanoparticles which transfer information between cells - offer significant potential in detecting and treating disease. Exosomes are the most promising next generation natural nano-vehicles for targeted drug and gene delivery. These are nano-sized vesicles (with diameter 40-200 nm) derived from patients' own healthy cells with an exceptional ability to interact with cellular membranes.

They possess a unique property of "cell specific tropism" (target specific cells by displaying receptors in the membranes) towards the originated cells, which can be utilized as a delivery strategy to transport cargo consisting of drugs, proteins and micro RNAs. Since they are of biological origin containing natural lipid bilayers, immunogenicity and drug clearance from the body can be reduced and can easily cross the blood-brain barrier, which are beneficial for designing personalized therapeutic approaches.

Recommended videos:

Injectable Nanoparticle Generator (iNG):

Injectable Nanoparticle Generator is the first of its kind new drug delivery strategy developed by the scientists at Houston Methodist Research Institute in Texas. It consists of a Dox loaded polymer made up of multiple strands enwrapped in a biodegradable nano-porous silicon material. When injected intravenously, due to natural tropism, they accumulate at the tumors, where the silicon material degrades slowly releasing the drug polymeric strands.

These strands spontaneously form nanoparticles that are then taken up by the cancer cells. The acidic environment inside the cancer cells triggers the polymeric strands to release the drug. This kind of novel approach helps the drug to cross multiple biological barriers in order to achieve targeted therapy.

Recommended videos:

Nano-terminators:

Researchers at North Carolina State University developed biodegradable liquid metal nano-terminators that are drug loaded nanodroplets made of a liquid-phase eutectic gallium-indium core and a thiolated polymeric shell and thiolated hyaluronic acid) to target cancer cells. These droplets when injected into the blood stream get absorbed into the tumors and release the drug by dissolving the liquid metal due to the presence of highly acidic tumor environment.

Recommended videos:

https://www.bing.com/videos/search?q=Nano-terminators&&view=detail&mid=CD902CE437EA9BA08149CD902CE437EA9BA08149&&FORM=VRDGAR

Dendrimers:

Dendrimers are nano-polymers with a well-defined structure, which is different from linear polymer molecules. It has a core at the center consisting of an atom or a molecule and branches emerging from core comprises of repeated units having one branch junction, called as generations, and many terminal groups also at the surface of the generations. Hence the framework of dendrimer can be controlled to make a good carrier.

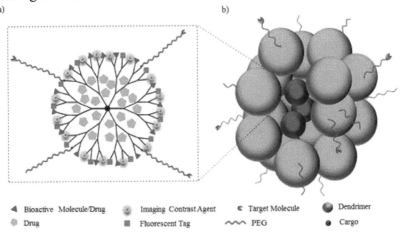

a) b)

| ◀ Bioactive Molecule/Drug | Imaging Contrast Agent | ∈ Target Molecule | Dendrimer |
| Drug | Fluorescent Tag | ∿ PEG | Cargo |

Recommended videos:

*https://www.bing.com/videos/search?q=Dendrimers&&view=detail&mid=62A06
1B4BB1888DCC35C62A061B4BB1888DCC35C&&FORM=VRDGAR*

*https://www.bing.com/videos/search?q=Dendrimers&&view=detail&mid=D7047
EAB7CB578902BF3D7047EAB7CB578902BF3&&FORM=VRDGAR*

*https://www.bing.com/videos/search?q=Dendrimers&&view=detail&mid=C6F69
873AE5A11A98DA3C6F69873AE5A11A98DA3&&FORM=VDRVRV*

Liposomes:

The distinctive feature of liposomes is its ability to compartmentalize and solubilize both hydrophilic and hydrophobic materials. For several years now, significant research has been underway on liposomal delivery of different drugs. One such example is ATP-mediated liposomal drug delivery. The method can be likened to keeping a cancer-killing bomb and its detonator separate until they are inside a cancer cell, where they then combine to destroy the cell.

Recommended videos:

*https://www.bing.com/videos/search?q=Liposomes&&view=detail&mid=241435
17CC25FD4A802024143517CC25FD4A8020&&FORM=VRDGAR*

*https://www.bing.com/videos/search?q=Liposomes&&view=detail&mid=3C7FB
D6F9A946B7431043C7FBD6F9A946B743104&&FORM=VRDGAR*

*https://www.bing.com/videos/search?q=Liposomes&&view=detail&mid=FF0B0
8C78C4DD2381F71FF0B08C78C4DD2381F71&&FORM=VDRVRV*

Niosomes:

Niosomes are also known as nonionic surfactant vesicles with sizes ranging from 20 nm to around 50 µm. They are formed from self-assembly of hydrated synthetic nonionic surfactant monomers and are capable of entrapping a variety of drugs. Niosomes have been evaluated as an alternative to liposomes in order to overcome their stability problems. Their unique structure helps to encapsulate both hydrophilic and lipophilic drug substances. Entrapment efficiency increases with increase in the concentration and lipophilicity of the surfactant used. In cancer therapy active targeting involves recognition of specific site (targeted site) and delivery to that place. It was achieved by coupling drug or ligand with the delivery system, which can interact with specific receptor at the targeted site and can be released from the carrier to commence destructive action on cell.

Recommended videos:

https://www.bing.com/videos/search?q=Niosomes&&view=detail&mid=7CAAA3 24D4BEB56B7EEE7CAAA324D4BEB56B7EEE&&FORM=VRDGAR

https://slideplayer.com/slide/5698414/

Carbon nanotubes (CNTs):

Carbon nanotubes offer several advantages for delivering drugs to specific locations inside the body which suggest that they may provide an improved result over nanoparticles. They have a larger inner volume which allows more drug molecules to be encapsulated, and this volume is more easily accessible because the end caps can be easily removed. And they have distinct inner and outer surfaces for functionalization.

All that is quite rare and that is why it brought our attention. Carbon nanotubes can also be chemically modified to carry a variety of molecules such as drugs, DNA, proteins, peptides, targeting ligands etc. (and often in combinations) into cells - which makes them suitable candidates for targeted delivery applications. Are there any other potential medical applications for this unique material? Most likely, yes, considering the unique chemical and physical properties of this material. More work must be done widening the areas of research. There are numerous companies and educational institutions that lead the way, and more and more money is poured in that technology in recognition of its importance.

Recommended videos:

https://www.bing.com/videos/search?q=Carbon+nanotubes+&&view=detail&mid=49687B0CCE1ED253FC6B49687B0CCE1ED253FC6B&&FORM=VRDGAR

https://www.bing.com/videos/search?q=Carbon+nanotubes+&&view=detail&mid=4B7208BFCB8CC32A2A984B7208BFCB8CC32A2A98&&FORM=VRDGAR

https://www.bing.com/videos/search?q=Carbon+nanotubes+&&view=detail&mid=62108B94895A4160901F62108B94895A4160901F&&FORM=VRDGAR

Graphene:

Another novel carbon nanomaterial, graphene, is beginning to be researched for nanomedicine applications as well. For instance, researchers have developed a simple method to thermally ablate highly resistant cancer cells using targeted biodegradable graphene nanoparticles. Scientists also have used graphene to target and neutralize cancer stem cells while not harming other cells. This new development opens the possibility of preventing or treating a broad range of cancers, using a non-toxic material. Other novel applications involve graphene-based sensors to detect cancer cells. There are some other candidates for medical research including graphene and its properties.

Graphene is an allotrope of carbon in the form of a two-dimensional, atomic-scale, honey-comb lattice in which one atom forms each vertex. It is the basic structural element of other allotropes, including graphite, charcoal, carbon nanotubes and fullerenes.

Recommended videos:

https://www.bing.com/videos/search?q=Graphene+&&view=detail&mid=388A6
1F7278B38C48C71388A61F7278B38C48C71&&FORM=VRDGAR

https://www.bing.com/videos/search?q=Graphene+&&view=detail&mid=BB9E
EAE47A20BD585E68BB9EEAE47A20BD585E68&&FORM=VRDGAR
https://www.bing.com/videos/search?q=Graphene+&&view=detail&mid=61608
C366FAD9327BEE561608C366FAD9327BEE5&&FORM=VDRVRV

https://www.bing.com/videos/search?q=Graphene++bullet-
resistant+vest&&view=detail&mid=68BDBDFEA40DC6050BF568BDBDFEA4
0DC6050BF5&&FORM=VRDGAR

https://www.bing.com/videos/search?q=Graphene++bullet-
resistant+vest&&view=detail&mid=015154B6B83C0BF02A82015154B6B83C0
BF02A82&&FORM=VDRVRV

Conclusion:

Nanotechnologies have enabled novel solutions for the treatment of various diseases. Nanodrug delivery systems present some advantages than conventional drug delivery systems such as high cellular uptake and reduced side effects. Development of drug delivery systems by using nanotechnology particularly for cancer treatment is making revolutionary changes in treatment methods and handling side effects of chemotherapy. Furthermore, nanotechnology allows for selective targeting of disease and infection containing cells and malfunctioned cells.

Recommended videos:

https://www.bing.com/videos/search?q=Nanotechnologies&&view=detail&mid=BAC0FF
2D2B4ADBB0018ABAC0FF2D2B4ADBB0018A&&FORM=VRDGAR

https://www.bing.com/videos/search?q=Nanotechnologies&&view=detail&mid=3439ABB
144AB8D6ED5923439ABB144AB8D6ED592&&FORM=VDRVRV

https://www.bing.com/videos/search?q=Nanotechnologies&&view=detail&mid=881CA78
EB2E5E007A521881CA78EB2E5E007A521&&FORM=VDRVRV

6.0   Inorganic Nanomaterials

Introduction:

Gold and silver nanoparticles have been used throughout the history for aesthetic and medical purposes. Mixtures of gold salts with molten glass were used by medieval artisans to produce tiny gold colloids exhibiting a ruby color. This was used to add color to ceramics and pottery. We will examine here the synthesis, properties, and applications of metal nanoparticles, nanowires, and quantum dots.

Recommended videos:

https://www.bing.com/videos/search?q=Inorganic+Nanomaterials&&view=detail&mid=A05B9F8E9FDAE11DB4D8A05B9F8E9FDAE11DB4D8&&FORM=VRDGAR

https://www.bing.com/videos/search?q=Inorganic+Nanomaterials&&view=detail&mid=C962D9554A55CFC601F2C962D9554A55CFC601F2&&FORM=VRDGAR

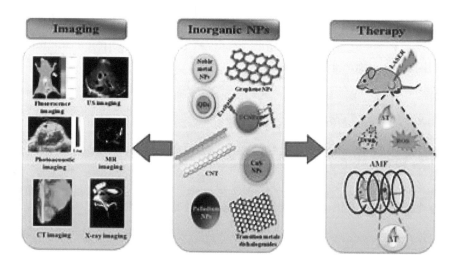

Physical and Chemical Properties of Metal Nanoparticles:

Nanoparticles are particles between 1 and 100 nanometers in size. In nanotechnology, a particle is defined as a small object that behaves as a whole unit with respect to its transport and properties. Particles are further classified according to diameter. While bulk materials have constant physical properties regardless of size, the size of a nanoparticle dictates its physical and chemical properties. Thus, the properties of a material change as its size approaches nanoscale proportions and as the percentage of atoms at the surface of a material becomes significant.

The physical and chemical properties of metallic nanoparticles are considerably different than for bulk metals. Metal particles exhibit lower melting points, higher surface areas, specific optical properties, and unusual mechanical strength. Due to these unique properties, metal nanoparticles often are chosen for specific tasks in industrial applications.

Nano-scale effects on properties

Properties	Examples
Catalytic	Better catalytic efficiency through higher surface-to-volume ratio
Electrical	Increased electrical conductivity in ceramics and magnetic nanocomposites, increased electric resistance in metals
Magnetic	Increased magnetic coercivity up to a critical grain size, superparamagnetic behaviour
Mechanical	Improved hardness and toughness of metals and alloys, ductility and superplasticity of ceramic
Optical	Spectral shift of optical absorbtion and fluorescence properties, increased quantum efficiency of semiconductor crystals
Sterical	Increased selectivity, hollow spheres for specific drug transportation and controlled release
Biological	Increased permeability through biological barriers (membranes, blood-brain barrier, etc.), improved biocompatibility

Metallic nanoparticles have fascinated scientist for over a century and are now heavily utilized in biomedical sciences and engineering. They are a focus of interest because of their properties and the huge potential in nanotechnology. Today these materials can be synthesized and modified with various chemical functional groups which allow them to be conjugated with antibodies, ligands, and drugs of interest and thus opening a wide range of potential applications in biotechnology, magnetic separation, and preconcentration of target analytes, targeted drug delivery, and vehicles for gene and drug delivery and more importantly diagnostic imaging.

Metal nanoparticles exhibit interesting linear optical properties that are analogous to molecular systems in effect but arise from a different physical process. Plasmon resonance absorption, which gives metal nanoparticles their characteristic color, is highly dependent on particle shape, and interparticle interaction. The electrochemical template synthesis method has provided a convenient method for studying the optical consequences of metal nanoparticle pair interactions. Using simple electrodeposition techniques, it is possible to control the distance between pair members, and thus cause systematic shifts in the plasmon resonance maximum. By controlling the relative sizes of the pair members, it is possible to enhance the second order NLO activity of small nanoparticle systems.

Nanoparticles consist of three layers: the surface layer, the shell layer, and the core. The surface layer usually consists of a variety of molecules such as metal ion, surfactants, and polymers. Nanoparticles may contain a single material or comprise of a combination of several materials. Nanoparticles can exist as suspensions, colloids, or dispersed aerosols depending on their chemical and electromagnetic properties. And, this is another reason to study it.

The properties of nanoparticles are hugely dependent on their size. For instance, copper nanoparticles that are smaller than 50 nm are presenting a super hard material and do not exhibit the properties of malleability or ductility of bulk copper. Other changes that are dependent on the size of nanoparticles are superparamagnetism exhibited by magnetic materials, quantum confinement by semiconductor Q-particles, and surface plasmon resonance in some metal particles.

Research has also demonstrated that absorption of solar radiation in photovoltaic cells is much higher in nanoparticles than it is in thin films of continuous sheets of bulk material. This is because nanoparticles are smaller and can absorb greater amount of solar radiation and faster. Nanoparticles are circular and that works to their advantage when absorbing and releasing the energy. Think of a layer of pebbles versus a solid piece of concrete.

Nanoparticles exhibit enhanced diffusion at elevated temperatures due to their high surface area to volume ratio. This property of nanoparticles allows sintering to take place at lower temperatures than in the case of larger particles. This could be a very valuable asset in many developments. While this diffusion property exhibited by nanoparticles may not affect the density of the product, it can lead to agglomeration.

Recommended videos:

https://www.bing.com/videos/search?q=Physical+and+Chemical+Properties+of+Metal+Nanoparticles&&view=detail&mid=14A2BA43B843B6FF3D7314A2BA43B843B6FF3D73&&FORM=VRDGAR

https://slideplayer.com/slide/6820237/

https://www.bing.com/videos/search?q=Physical+and+Chemical+Properties+of+Metal+Nanoparticles&&view=detail&mid=74A51854936080901A5174A51854936080901A51&&FORM=VRDGAR

Band Theory and Quantum Confinement:

Having a few atoms in a material leads to other exceptional properties associated with nanoparticles. In bulk materials, atomic orbitals overlap constructing bands. In nanoscale metals orbitals are not continuous, they are spaced apart each with representative energy. This means the band gap of metal nanoparticles could be turned on and off just by altering nanoparticle diameter.

As the particle size decreases, the electron becomes more confined in the particle. In result, valence bands (filled atomic orbitals) and conduction bands (empty atomic orbitals) break into quantized energy level. When the atomic orbitals of nanoparticles become discrete or quantized, quantum confinement occurs. **Quantum confinement** is change of electronic and optical properties when the material sampled is of sufficiently small size - typically 10 nanometers or less. The bandgap increases as the size of the nanostructure decreases. **Quantum confined** structure is one in which the motion of the carriers (electron and hole) are **confined** in one or more directions by potential barriers. Based on the **confinement** direction, a **quantum confined** structure will be classified into three categories as **quantum** well, **quantum** wire and **quantum** dots.

Quantum confinement is responsible for the **increase of energy difference between energy states and band gap**, a phenomenon tightly related to the optical and electronic properties of the materials. Quantum confinement can be observed once the diameter of a material is of the same magnitude as the de Broglie wavelength of the electron wave function. (De Broglie wavelength. The wavelength $\gamma = h/p$ associated with a beam of particles (or with a single particle) of momentum p; $h = 6.626 \times 1034$ joule-second is Planck's constant).

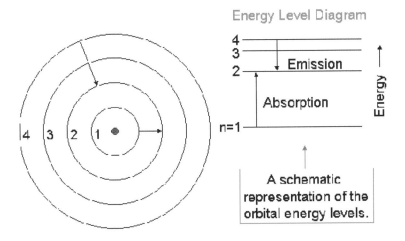

A schematic representation of the orbital energy levels.

Recommended videos:

https://www.bing.com/videos/search?q=Quantum+Confinement+Explanation&&view=detail&mid=1342C54B8563893B4F821342C54B8563893B4F82&&FORM=VRDGAR

https://www.bing.com/videos/search?q=Band+Theory+and+Quantum+Confinement&&view=detail&mid=D94269CF227306DC06BCD94269CF227306DC06BC&&FORM=VRDGAR

https://www.bing.com/videos/search?q=Band+Theory+and+Quantum+Confinement&&view=detail&mid=3C5BEB156D686CAA46433C5BEB156D686CAA4643&&FORM=VRDGAR

https://www.bing.com/videos/search?q=quantum+confinement&&view=detail&mid=C1EB800A0578AA7C8B75C1EB800A0578AA7C8B75&&FORM=VRDGAR

Surface Plasmon Resonance and Optical Properties:

The exceptional band structure of metal nanoparticles inspires very interesting light-matter interactions. When a metal particle is exposed to light, the electric field component of the electromagnetic wave induces collective, coherent oscillations of electrons in the nanoparticle. This is referred to as surface plasmon resonance (SPR). During this phenomenon, the electric field displaces electron clouds.

Surface plasmon resonance (SPR) is the resonant oscillation of conduction electrons at the interface between a negative and positive permittivity material stimulated by incident light. The resonance condition is established when the frequency of incident photons matches the natural frequency of surface electrons oscillating against the restoring force of positive nuclei. The unique **optical properties** of **plasmonic** nanoparticles have been observed for thousands of years. Since ancient times artists have used colloidal nanoparticles of gold, silver, and copper to give color to pottery and stained glass.

Plasmonic nanoparticles are particles whose electron density can couple with electromagnetic radiation of wavelengths that are far larger than the particle due to the nature of the dielectric-metal interface between the medium and the particles: unlike in a pure metal where there is a maximum limit on what size wavelength can be effectively coupled based on the material size.

Surface Plasmon Resonance Application

Resonance conditions
- Matching
  Electron Freq and Light Freq
- Depending on Electronic Properties
  Surface and Surrounding Medium
- Angle of Incidence

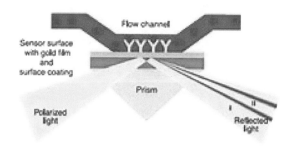

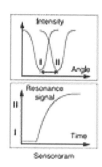

Plasmonic metamaterials are metamaterials that exploit surface plasmons to achieve optical properties not seen in nature. Plasmons are produced from the interaction of light with metal-dielectric materials. Under specific conditions, the incident light couples with the surface plasmons to create self-sustaining, propagating electromagnetic waves known as surface plasmon polaritons (SPPs). Surface plasmons (SPs) are coherent delocalized electron oscillations that naturally exist at the interface between any two materials where the real part of the dielectric function changes signs across the interface (e.g. a metal-dielectric interface, such as a metal sheet in air).

Surface plasmon resonance is an optical technique utilized for detecting molecular interactions. Binding of a mobile molecule (analyte) to a molecule immobilized on a thin metal film (ligand) changes the refractive index of the film. The angle of extinction of light, reflected after polarized light impinges upon the film, is altered, monitored as a change in detector position for the dip in reflected intensity (the surface plasmon resonance phenomenon). Because the method strictly detects mass, there is no need to label the interacting components, thus eliminating possible changes of their molecular properties. We have utilized surface plasmon resonance to study the interaction of proteins of hair cells.

Hence, the development of **optical** biosensors based on **optical properties** of noble metal nanoparticles using **Surface Plasmon Resonance** was considered and it brought the results. Multiple biosensors were developed already and employed throughout the medical industry. More is under investigation and the results are not that far away. **Surface plasmon resonance** (SPR) is a powerful technique to retrieve information on **optical properties** of biomaterial and nanomaterials.

Recommended videos:

https://www.bing.com/videos/search?q=Surface+Plasmon+Resonance+and+Opt ical+Properties&&view=detail&mid=03DA3164CCFCC985E16703DA3164CCF CC985E167&&FORM=VDRVRV

https://www.bing.com/videos/search?q=Surface+Plasmon+Resonance+and+Opt ical+Properties&&view=detail&mid=8FB2A1CD1B67056C71578FB2A1CD1B6 7056C7157&&FORM=VDRVRV

https://www.bing.com/videos/search?q=Surface+Plasmon+Resonance+and+Opt ical+Properties&&view=detail&mid=5C9D20ADC577BB224F0B5C9D20ADC5 77BB224F0B&&FORM=VRDGAR

https://www.bing.com/videos/search?q=Surface+Plasmon+Resonance+and+Opt ical+Properties&&view=detail&mid=E843E6A532AAEABECC3BE843E6A532 AAEABECC3B&&FORM=VDRVRV

Metal Nanoparticles Synthesis:

Microbial synthesis of metal nanoparticles depends up on the localization of the reductive components of the cell. When the cell wall reductive enzymes or soluble secreted enzymes are involved in the reductive process of metal ions then it is obvious to find the metal nanoparticles extracellularly. Many scientists believe that all metal nanoparticles could be synthesized eventually, and they work very hard trying to prove the point.

Metal nanoparticles research has recently become the focus of intense work due to their unusual properties compared to bulk **metal**. After the pioneer work on preparation of Pt, Pd, Rh and Ir **nanoparticles**, done using the water-in-oil microemulsion method, several other **metallic nanoparticles** were **synthesized** and are going to be made commercial. The scientific research in the area of metal nanoparticles is never done. So far, we have scratched only the surface and that is not even the beginning. This is only the beginning of the beginning. There are more and more opportunities seen on the horizon.

SYNTHESIS METHODS:

A.   CHEMICAL METHODS

A.1.   Chemical reduction of metal salts;
       A.1.1. The alcohol reduction process;
       A.1.2. The polyol process;
A.2.   Microemulsions;
A.3.   Thermal decomposition of metal salts;
A.4.   Electrochemical synthesis;

B.   PHYSICAL METHODS

B.1.   Exploding wire technique;
B.2.   Plasma;
B.3.   Chemical vapor deposition;
B.4.   Microwave irradiation;
B.5.   Pulsed laser ablation;
B.6.   Supercritical fluids;
B.6.   Sonochemical reduction;

Sonochemistry is the application of ultrasound to chemical reactions and processes. The mechanism causing sonochemical effects in liquids is the phenomenon of acoustic cavitation. Some very interesting developments resulted from that.

B.7.   Gamma radiation;

Gamma ray (also called gamma radiation), denoted by the lower-case Greek letter gamma, is extremely high-frequency electromagnetic radiation and therefore consists of high-energy photons. Paul Villard, a French chemist and physicist, discovered gamma radiation in 1900 while studying radiation emitted by radium. In 1903, Ernest Rutherford named this radiation gamma rays.

Recommended videos:

https://www.bing.com/videos/search?q=Metal+Nanoparticles+Synthesis&&view =detail&mid=208F3410DBB032CC3B54208F3410DBB032CC3B54&&FORM =VRDGAR

https://www.bing.com/videos/search?q=Metal+Nanoparticles+Synthesis&&view =detail&mid=66B94EC177FEBAF1F60566B94EC177FEBAF1F605&&FORM =VRDGAR

Goals and Problems in Metallic Nanoparticles Synthesis:

Ideally, metallic nanoparticles should be prepared by a method which:
- ✓ Is reproducible;
- ✓ may control the shape of the particles;
- ✓ yields monodisperse metallic nanoparticles;
- ✓ is easy, cheap;
- ✓ use fewer toxic precursors: in water or more environmentally benign solvents (e.g. Ethanol);
- ✓ use the least number of reagents;
- ✓ use a reaction temperature close to room temperature with as few synthetic steps as possible (one-pot reaction) minimizing the quantities of generated by-products and waste.

Recommended videos:

https://www.bing.com/videos/search?q=Goals+and+Problems+in+Metallic+Nan oparticles+Synthesis&&view=detail&mid=12CF8AD75B6FF01DD0CB12CF8A D75B6FF01DD0CB&&FORM=VDRVRV

Metal Nanoparticles in Medicine:

Colloidal gold nanoparticles have been employed for a relatively long time for the treatment of diseases including cancer, rheumatoid arthritis, multiple sclerosis, and neurodegenerative conditions such as Alzheimer's disease. The advantages of using gold nanoparticles in medical applications include, ease of preparation, range of sizes, good biocompatibility, ease of functionalizing, and ability to conjugate with other biomolecules without altering their biological properties.

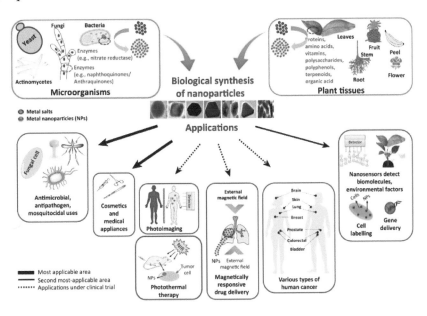

Recent years have witnessed tremendous applications of metal NPs in diverse fields of research including material science, biotechnology, biomedical engineering, targeted drug delivery, environmental, etc. In addition, the potential of metal NPs in separation science has been already well recognized due to their easy synthesis procedure, large s/v ratio, controllable particle size, narrow size distribution, extraordinary biocompatibility, molecular detection properties, etc.

Therapies that involve the manipulation of individual genes, or the molecular pathways that influence their expression, are increasingly being investigated as an option for treating diseases. One highly sought goal in this field is the ability to tailor treatments according to the genetic make-up of individual patients. This creates a need for tools that help scientists experiment and develop such treatments.

Imagine, for example, being able to stretch out a section of DNA like a strand of spaghetti, so you can examine or operate on it, or building nanorobots that can "walk" and carry out repairs inside cell components. Nanotechnology is bringing that scientific dream closer to reality.

Scientists are discovering that protein-based drugs are very useful because they can be programmed to deliver specific signals to cells. But the problem with conventional delivery of such drugs is that the body breaks most of them down before they reach their destination.

But what if it were possible to produce such drugs *in situ*, right at the target site (in the original place)? Well, in a recent issue of *Nano Letters*, researchers at Massachusetts Institute of Technology (MIT) in the US show how it may be possible to do just that. In their proof of principle study, they demonstrate the feasibility of self-assembling "nano-factories" that make protein compounds, on demand, at target sites. So far, they have tested the idea in mice, by creating nanoparticles programmed to produce either green fluorescent protein (GFP) or luciferase exposed to UV light.

The MIT team came up with the idea while trying to find a way to attack metastatic tumors that grow from cancer cells that have migrated from the original site to other parts of the body. Over 90% of cancer deaths are due to metastatic cancer. They are now working on nanoparticles that can synthesize potential cancer drugs, and also on other ways to switch them on.

Nanofibers are fibers with diameters of less than 1,000 nm. Medical applications include special materials for wound dressings and surgical textiles, materials used in implants, tissue engineering and artificial organ components. Nanofibers made of carbon also hold promise for medical imaging and precise scientific measurement tools. But there are huge challenges to overcome, one of the main ones being how to make them consistently of the correct size. Historically, this has been costly and time-consuming.

Drugs may be also linked to the surface. Metal nanoparticles have a special property of high surface area to volume ration, which facilitates attachment of various functional groups, allowing the nanoparticles to bind to tumor cells. Once they are at the target site, the drug payload is released from the nanoparticle by diffusion, swelling, erosion, or degradation.

Recommended videos:

https://www.bing.com/videos/search?q=Metal+Nanoparticles+in+Medicine&&view=detail&mid=9EDD3A985686B0DFFA889EDD3A985686B0DFFA88&&FORM=VRDGAR

https://www.bing.com/videos/search?q=Metal+Nanoparticles+in+Medicine&&view=detail&mid=BC67C7AE64C3830B9A06BC67C7AE64C3830B9A06&&FORM=VDRVRV

https://www.bing.com/videos/search?q=Metal+Nanoparticles+in+Medicine&&view=detail&mid=43F20333515C87B4DD1C43F20333515C87B4DD1C&&FORM=VDRVRV

Quantum Dots:

Quantum dots are nanoscale particle of semiconducting material that can be embedded in cells or organisms for various experimental purposes, such as labeling proteins.

Quantum dots (QD) are nanoscale semiconductor devices that tightly confine either electrons or electron holes in all three spatial dimensions. They can be made via several possible routes including colloidal synthesis, plasma synthesis, or mechanical fabrication.

A quantum dot gets its name because it's a tiny speck of matter so small that it's effectively concentrated into a single point (in other words, it's **zero-dimensional**). As a result, the particles inside it that carry electricity (electrons and holes, which are places that are missing electrons) are trapped ("constrained") and have well-defined energy levels according to the laws of quantum theory, a bit like individual atoms. Tiny really does mean *tiny*: quantum dots are crystals a few nanometers wide, so they're typically a few dozen atoms across and contain anything from perhaps a hundred to a few thousand atoms. They're made from a **semiconductor** such as silicon (a material that's neither really a conductor nor an insulator but can be chemically treated so it behaves like either). And although they're crystals, they behave more like individual atoms - hence the nickname artificial atoms.

Quantum dots can be precisely controlled to do all kinds of useful things. School-level physics tells us that if you give an atom energy, you can "excite" it. You can boost an electron inside it to a higher energy level. When the electron returns to a lower level, the atom emits a photon of light with the same energy that the atom originally absorbed. The color (wavelength and frequency) of light an atom emits depends on what the atom is; iron looks green when you excite its atoms by holding them in a hot flame, while sodium looks yellow, and that's because of the way their energy levels are arranged. The rule is that different atoms give out different colors of light. All this is possible because the energy levels in atoms have set values; in other words, they are **quantized**.

So far, quantum dots have attracted most interest because of their interesting optical properties. They're being used for all sorts of applications where precise control of colored light is important. In one simple and relatively trivial application, a thin filter made of quantum dots has been developed so it can be fitted on top of a fluorescent or LED lamp and convert its light from a blueish color to a warmer, redder, more attractive shade quite similar to the light produced by old-fashioned incandescent lamps. Quantum dots can also be used instead of pigments and dyes. Embedded in other materials, they absorb incoming light of one color and give out light of an entirely different color; they're brighter and more controllable than organic dyes (artificial dyes made from synthetic chemicals).

Quantum dots are also finding important medical applications, including potential cancer treatments. Dots can be designed so they accumulate in specific parts of the body and then deliver anti-cancer drugs bound to them. Their big advantage is that they can be targeted at single organs, such as the liver, much more precisely than conventional drugs, so reducing the unpleasant side effects that are characteristic of untargeted, traditional chemotherapy. Could that apply to some other sicknesses and medications as well?

Quantum dots are also being used in place of organic dyes in biological research. For example, they can be used like nanoscopic light bulbs to light up and color specific cells that need to be studied under a microscope. They're also being tested as sensors for chemical and biological warfare agents such as anthrax. Unlike organic dyes, which operate over a limited range of colors and degrade relatively quickly, quantum dyes are very bright, can be made to produce any color of visible light, and theoretically last indefinitely (they are said to be photostable).

Recommended videos:

*https://www.bing.com/videos/search?q=Quantum+Dots&&view=detail&mid=B2
F55BFB84E95BBD2781B2F55BFB84E95BBD2781&&FORM=VRDGAR*

*https://www.bing.com/videos/search?q=Quantum+Dots&&view=detail&mid=23
67B4BD5677148D34FC2367B4BD5677148D34FC&&FORM=VRDGAR*

*https://www.bing.com/videos/search?q=Quantum+Dots&&view=detail&mid=22
AB85C1F359DD68EBDF22AB85C1F359DD68EBDF&&FORM=VDRVRV*

Quantum Dot Synthesis:

Quantum dots made from aqueous **synthesis** is especially attractive for biological application due to their compatibility with water. Also compared to the organic-based **synthesis**, aqueous **synthesis** is cheaper, less toxic and more environmentally friendly.

Quantum dots are nicknamed **"artificial atoms"**. When photons are pumped into a semiconductor, electrons are excited into the conduction band, leaving behind holes in the valence band. Binding the electrons with their hole counterparts result in bounded electron-hole pairs, or excitons. **Quantum dots** can best be described as false **atoms**.

The essential elements for the synthesis of quantum dots involve combining an appropriate metallic, or organometallic precursors such as zinc, cadmium or mercury species with a corresponding chalcogen precursor such. as **sulfur, selenium or tellurium species in. coordinating organic solvent at high temperatures**.

Recommended videos:

https://www.bing.com/videos/search?q=Quantum+Dot+Synthesis&&view=detail&mid=46A7FA6CA35AF7A6057146A7FA6CA35AF7A60571&&FORM=VRDGAR

https://www.bing.com/videos/search?q=Quantum+Dot+Synthesis&&view=detail&mid=EA1A20918A256EE0A357EA1A20918A256EE0A357&&FORM=VDRVRV

Optical Properties of Quantum Dots:

The **optical properties of quantum dots** are known to vary between different types and can be predicted by certain factors. The material that the **quantum dot** is constructed from plays a role in determining the intrinsic energy signature of the **particle**, but the most important factor that affects the **optical properties** is the size of the **dots**.

Quantum dots can emit any color of light from the same material by changing the dot size. They have bright, pure colors along that can emit a rainbow of colors coupled with their high efficiencies, longer lifetimes and high extinction coefficient. The high extinction coefficient of a quantum dot makes it perfect for optical uses. Quantum dots of very high quality can be ideal for applications in optical encoding and multiplexing due to their narrow emission spectra and wide excitation profile.

Wavelengths of light emitted by a **quantum dot** depends on its size. As the size decreases, the wavelength emitted also shortens, and moves toward the blue end of the visual **electromagnetic** spectrum 2. Conversely, increasing the **dot** size lengthens the wavelengths emitted, moving toward the red end.

Recommended videos:

https://www.bing.com/videos/search?q=Optical+Properties+of+Quantum+Dots &&view=detail&mid=3FDBC150D7A67F701CE03FDBC150D7A67F701CE0& &FORM=VRDGAR

https://www.bing.com/videos/search?q=Optical+Properties+of+Quantum+Dots &&view=detail&mid=2CD8CFD4422D789235652CD8CFD4422D78923565&& FORM=VRDGAR

Single Electron Devices:

It is possible for quantum dots to be used to create single electron devices. The single electron transistor (SET) is a new type of switching device using controlled electron tunneling to switch current on and off. Tunneling is used to put a charge in a quantum dot surrounded by an insulated barrier. Single-electron devices are promising, as new nanoscale devices, because they can control the motion of a single electron.

A gate voltage is used to control the opening and closing of the SET, or in other words, it controls the one-by-one electron transfer. When there is no bias on any electrode, electrons in the system do not have enough energy to tunnel through the junctions.

Recommended videos:

https://www.bing.com/videos/search?q=Single+Electron+Devices&&view=detail&mid=16 FBD2708C6F054D9A0916FBD2708C6F054D9A09&&FORM=VRDGAR

https://slideplayer.com/slide/2453763/

https://www.bing.com/videos/search?q=Single+Electron+Devices&&view=detail&mid=B CF412BD2626F008D66EBCF412BD2626F008D66E&&FORM=VRDGAR

Quantum Dots in Medicine:

Doxorubicin, an anthracycline antibiotic that is widely used in chemotherapy, can be immobilized onto quantum dots in order to improve and control the kinetics of drug release. In addition, quantum dots can act as delivery vehicles for small interfering RNAs, which are powerful tools for silencing gene expression. **Quantum dots** enable researchers to study cell processes at the level of a single molecule and may significantly improve the diagnosis and **treatment** of diseases such as cancers.

Cadmium selenide nanoparticles, in the form of quantum dots, are employed in the detection of cancer tumors because they emit bright energy when exposed to ultraviolet light. This facilitates easy removal of tumors since quantum dots possess size-tunable light emissions. They are often used in conjunction with magnetic resonance imaging to produce exceptional images of tumor sites. The results, as compared to dyes, are much brighter and need only one light source. Quantum dots produce a higher contrast image at a lower cost than organic dyes routinely employed as contrast media.

Recommended videos:

https://www.bing.com/videos/search?q=Quantum+Dots+in+Medicine&&view=detail&mid=3BD10F6BB4B21659F7C83BD10F6BB4B21659F7C8&&FORM=VRDGAR

https://www.bing.com/videos/search?q=Quantum+Dots+in+Medicine&&view=detail&mid=99976EBD32E6F2EBF2DF99976EBD32E6F2EBF2DF&&FORM=VRDGAR

Metal Nanowires:

A **nanowire** is a nanostructure, with the diameter of the order of a nanometer (10^{-9} meters). It can also be defined as the ratio of the length to width being greater than 1000. Alternatively, nanowires can be defined as structures that have a thickness or diameter constrained to tens of nanometers or less and an unconstrained length. At these scales, quantum mechanical effects are important - which coined the term "quantum wires". Many different types of nanowires exist, including superconducting, metallic, semiconducting (silicon wires) and insulating. Molecular nanowires are composed of repeating molecular units either organic (e.g. DNA) or inorganic.

Typical nanowires exhibit aspect ratios (length-to-width ratio) of 1000 or more. As such they are often referred to as one-dimensional (1-D) materials. Nanowires have many interesting properties that are not seen in bulk or 3-D (three-dimensional) materials. This is because electrons in nanowires are quantum confined laterally and thus occupy energy levels that are different from the traditional continuum of energy levels or bands found in bulk materials.

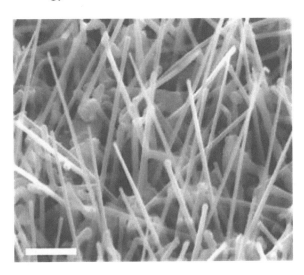

Metal Nanowires

Peculiar features of this quantum confinement exhibited by certain nanowires manifest themselves in discrete values of the electrical conductance. Such discrete values arise from a quantum mechanical restraint on the number of electrons that can travel through the wire at the nanometer scale. There are many applications where nanowires may become important in electronic, opto-electronic and nanoelectromechanical devices, as additives in advanced composites, for metallic interconnects in nanoscale quantum devices, as field-emitters and as leads for biomolecular nano-sensors.

Recommended videos:

https://www.bing.com/videos/search?q=Metal+Nanowires&&view=detail&mid= 37157F8630AB203C272237157F8630AB203C2722&&FORM=VRDGAR

https://www.bing.com/videos/search?q=Metal+Nanowires&&view=detail&mid= 7C7631B745888DCB89E27C7631B745888DCB89E2&&FORM=VRDGAR

Synthesis:

There are two basic approaches to synthesizing nanowires: top-down and bottom-up. A top-down approach reduces a large piece of material to small pieces, by various means such as lithography, milling or thermal oxidation. A bottom-up approach synthesizes the nanowire by combining constituent adatoms. Most synthesis techniques use a bottom-up approach. Initial synthesis via either method may often be followed by a nanowire thermal treatment step, often involving a form of self-limiting oxidation, to fine tune the size and aspect ratio of the structures. Nanowire production uses several common laboratory techniques, including suspension, electrochemical deposition, vapor deposition, and VLS growth. Ion track technology enables growing homogeneous and segmented nanowires down to 8 nm diameter. As nanowire oxidation rate is controlled by diameter, thermal oxidation steps are often applied to tune their morphology.

Recommended videos:

https://www.bing.com/videos/search?q=Metal+Nanowires+Synthesis&&view=detail&mid=C3748577E8D25D61A7D7C3748577E8D25D61A7D7&&FORM=VRDGAR

https://www.bing.com/videos/search?q=Metal+Nanowires+Synthesis&&view=detail&mid=CD32E0D676DC469830C6CD32E0D676DC469830C6&&FORM=VRDGAR

Physical Properties:

Conductivity of nanowires:

Some physical reasons predict that the conductivity of a nanowire will be much less than that of the corresponding bulk material. First, there is scattering from the wire boundaries, whose effect will be very significant whenever the wire width is below the free electron mean free path of the bulk material. In copper, for example, the mean free path is 40 nm. Copper nanowires less than 40 nm wide will shorten the mean free path to the wire width.

Nanowires also show other peculiar electrical properties due to their size. Unlike single wall carbon nanotubes, whose motion of electrons can fall under the regime of ballistic transport (the electrons can travel freely from one electrode to the other), nanowire conductivity is strongly influenced by edge effects. The edge effects come from atoms that lay at the nanowire surface and are not fully bonded to neighboring atoms like the atoms within the bulk of the nanowire. The unbonded atoms are often a source of defects within the nanowire and may cause the nanowire to conduct electricity more poorly than the bulk material. As a nanowire shrinks in size, the surface atoms become more numerous compared to the atoms within the nanowire, and edge effects become more significant.

Furthermore, the conductivity can undergo a quantization in energy: i.e. the energy of the electrons going through a nanowire can assume only discrete values. The conductivity is hence described as the sum of the transport by separate *channels* of different quantized energy levels. The thinner the wire is, the smaller the number of channels available to the transport of electrons.

This quantization has been demonstrated by measuring the conductivity of a nanowire suspended between two electrodes while pulling it: as its diameter reduces, its conductivity decreases in a stepwise fashion. The quantization of conductivity is more pronounced in semiconductors like Si than in metals, due to their lower electron density and lower effective mass. It can be observed in 25 nm wide silicon fins, and results in increased threshold voltage. In practical terms, this means that a MOSFET with such nanoscale silicon fins, when employed in digital applications, will need a higher gate (control) voltage to switch the transistor on.

Recommended videos:

https://www.bing.com/videos/search?q=Conductivity+of+nanowires&&view=detail&mid=835D149D0A7FC485D187835D149D0A7FC485D187&&FORM=VRDGAR

https://www.bing.com/videos/search?q=Conductivity+of+nanowires&&view=detail&mid=4189B9BE694AD19D7C434189B9BE694AD19D7C43&rvsmid=835D149D0A7FC485D187835D149D0A7FC485D187&FORM=VDRVRV

Welding Nanowires:

To incorporate nanowire technology into industrial applications, researchers in 2008 developed a method of welding nanowires together. A sacrificial metal nanowire, using the manipulators of a scanning electron microscope, is placed adjacent to the ends of the pieces to be joined. Then, an electric current is applied, which fuses the wire ends. The technique fuses wires as small as 10 nm.

For nanowires with diameters less than 10 nm, existing welding techniques will not be practical. That technique requires precise control of the heating mechanism and which may introduce the possibility of damage. Recently scientists discovered that single-crystalline ultrathin gold nanowires with diameters ~3–10 nm can be "cold-welded" together within seconds by mechanical contact alone, and under remarkably low applied pressures. High-resolution transmission electron microscopy and in situ measurements reveal that the welds are nearly perfect, with the same crystal orientation, strength and electrical conductivity as the rest of the nanowire. The high quality of the welds is attributed to the nanoscale sample dimensions, oriented-attachment mechanisms and mechanically assisted fast surface diffusion. Nanowire welds were also demonstrated between gold and silver, and silver nanowires (with diameters ~5–15 nm) at near room temperature, indicating that this technique may be generally applicable for ultrathin metallic nanowires. Combined with other nano- and microfabrication technologies, cold welding is anticipated to have multiple potential applications in the future bottom-up assembly of metallic one-dimensional nanostructures. The entire field of nanowires is presenting an outstanding opportunity for future research and development.

Recommended videos:

https://www.bing.com/videos/search?q=Welding+Nanowires&&view=detail&mid=4D2BF6BD2CE7FFDFD58A4D2BF6BD2CE7FFDFD58A&&FORM=VRDGAR

https://www.bing.com/videos/search?q=Welding+Nanowires&&view=detail&mid=6AE5534FC78C76CD95DA6AE5534FC78C76CD95DA&&FORM=VRDGAR

https://www.bing.com/videos/search?q=Welding+Nanowires&&view=detail&mid=EE3D75CCCB3CBD477C65EE3D75CCCB3CBD477C65&&FORM=VRDGAR

Mechanical Properties of Nanowires:

Investigation of mechanical properties of nanowires:

The study of nanowire mechanics has boomed since the advent of the Atomic Force Microscope (AFM), and associated technologies which have enabled direct study of the response of the nanowire to an applied electrical load. Specifically, a nanowire can be clamped from one end, and the free end displaced by an AFM tip. In this cantilever geometry, the height of the AFM is precisely known, and the force applied is precisely known. This allows for construction of a force vs. displacement curve, which can be converted to a stress vs. strain curve if the nanowire dimensions are known. From the stress-strain curve, the elastic constant known as the Young's Modulus (elasticity) can be derived, as well as the toughness, and degree of strain-hardening. Moreover, if this process occurs while simultaneously viewing the nanowire in a scanning electron microscope, the resulting mechanical properties can be directly correlated with the nanowire's microstructure.

Young's modulus of nanowires:

The elastic component of the stress-strain curve described by the Young's Modulus, has been reported for nanowires, however the modulus depends very strongly on the microstructure. Thus, a complete description of the modulus dependence on diameter is lacking. It could change in both directions depending on materials, and the conditions.

The modulus of elasticity (also known as the elastic modulus, the tensile modulus, or Young's modulus) is a number that measures an object or substance's resistance to being deformed elastically (i.e., non-permanently) when a force is applied to it. More experiments has to be done and the more we do, the more we know.

Yield strength of nanowires:

The plastic component of the stress strain curve (or more accurately the onset of plasticity) is described by the yield strength. The strength of a material is increased by decreasing the number of defects in the solid, which occurs naturally in nanomaterials where the volume of the solid is reduced. As a nanowire is shrunk to a single line of atoms, the strength should theoretically increase all the way to the molecular tensile strength. Gold nanowires have been described as "ultrahigh strength" due to the extreme increase in yield strength. This huge increase in yield is determined to be due to the lack of dislocations in the solid. Without dislocation motion, a "dislocation-starvation" mechanism is in operation. The material can accordingly experience huge stresses before dislocation motion is possible, and then begins to strain-harden. For these reasons, nanowires have been employed extensively in composites for increasing the overall strength of a material. Think of the rebars inserted into concrete for strength. Moreover, nanowires continue to be actively studied, with research aiming to translate enhanced mechanical properties to novel devices in the fields of MEMS or NEMS.

Recommended videos:

https://www.bing.com/videos/search?q=Properties+of+Nanowires&&view=detail&mid=ABB1235D5803EF02443BABB1235D5803EF02443B&&FORM=VRDGAR

Applications:

Electronic devices:

Nanowires can be used and are widely used for transistors. Transistors are employed broadly as fundamental building element in today's electronic circuits. We can find them in almost every electronic device. As predicted by Moore's law and with assistance of nanotechnology, the dimensions of transistors are quickly shrinking into nanoscale. One of the key challenges of building future nanoscale transistors is ensuring good gate control over the channel. Due to the high aspect ratio, if the gate dielectric is wrapped around the nanowire channel, we can get good control of channel electrostatic potential, thus turning the transistor on and off efficiently.

Due to the unique one-dimensional structure with remarkable optical properties, the nanowire also opens new opportunities for realizing high efficiency photovoltaic devices. Compared with its bulk counterparts, the nanowire solar cells are less sensitive to impurities due to bulk recombination, and thus silicon wafers with lower purity can be used to achieve acceptable efficiency, leading to a reduction on material consumption. Now, we can not only produce solar sells on nano scale but do it efficiently, and we are still not done yet. More and more highly efficiency photovoltaic devices are being researched and developed.

To create active electronic elements, the first key step was to chemically dope a semiconductor nanowire. This development was understood and employed in multiple processes used throughout the industry. This has already been done to individual nanowires to create **p-type** and **n-type** semiconductors.

The next step was to find a way to create a **p–n junction**, one of the simplest electronic devices. This was achieved in two ways. The first way was to physically cross a **p-type** wire over an **n-type** wire. The second method involved chemically doping a single wire with different dopants along the length. This method created a **p-n junction** with only one wire. This was and still is quite advanced. What else is there and we have not seen yet? More research is needed and more has to be done.

After building **p-n junctions** with nanowires, the next logical step was to build logic gates. By connecting several **p-n junctions** together, researchers have been able to create the basis of all logic circuits: the AND, OR, and NOT gates have all been built from semiconductor nanowire crossings. New developments are made quite often in this field that needs improvement in order to create less complicated and low-cost processes. It is moving forward with predictably high-speed research and development.

It is possible and may be even essential that semiconductor nanowire crossings will be important to the future of digital computing. Though there are other usages for nanowires beyond these, the only ones that truly take advantage of physics in the nanometer regime are electronic. Still, the medical field is not much behind and is gaining speed. In addition, nanowires are also being studied for employment as photon ballistic waveguides, as interconnects in quantum dot/quantum effect well photon logic arrays. Photons travel inside the tube, electrons travel on the outside shell. They can travel separately and together. What technical advantages does it offer us? We do not know yet, but we will.

We know that when two nanowires acting as photon waveguides cross each other, the juncture acts as a quantum dot. That is another area, avenue for research, and that is more than very promising. What happens at that juncture point? Can we control the outcome somehow? Do we always know the outcome there? This could be big, no, huge.

Conducting nanowires offer the possibility of connecting molecular-scale entities in a molecular computer. Dispersions of conducting nanowires in different polymers are being investigated for usage as transparent electrodes for flexible flat-screen displays. These are just some examples, but we are only on the beginning of understanding nanowires.

Because of their high Young's moduli, their use in mechanically enhancing composites is also being investigated. Since nanowires appear in bundles, they may be used as tribological (the study of friction, wear, lubrication, and the design of bearings; the science of interacting surfaces in relative motion) additives to improve friction characteristics and reliability of electronic transducers and actuators.

Because of their high aspect ratio, nanowires are also uniquely suited to dielectrophoretic manipulation, which offers a low-cost, bottom-up approach to integrating suspended dielectric metal oxide nanowires in electronic devices such as UV, water vapor, and ethanol sensors.

Dielectrophoretic (or DEP) is a phenomenon in which a force is exerted on a dielectric particle when it is subjected to a non-uniform electric field. This force does not require the particle to be charged. All particles exhibit dielectrophoretic activity in the presence of electric fields.

Recommended videos:

https://www.bing.com/videos/search?q=Applications+of+Nanowires&&view=detail&mid=B9A9F0213C36903BF57FB9A9F0213C36903BF57F&&FORM=VRDGAR

Nanowire Lasers:

The discovery and continued development of the laser has revolutionized both science and industry. The advent of miniaturized, semiconductor lasers has made this technology an integral part of everyday life. Exciting research continues with a new focus on nanowire lasers because of their great potential in the field of optoelectronics. This could be the biggest leap forward in the field of optoelectronics research and development. This field promises so much that the best scientific brains give it more and more attention as time progresses. What else is in store for us there?

Semiconductor nanowire lasers are nano-scaled lasers that can be embedded on chips and constitute an advance for computing and information processing applications. Nanowire lasers are coherent light sources as any other laser device, with the advantage of operating at the nanoscale. Built by molecular beam epitaxy, nanowire lasers offer the possibility for direct integration on silicon, and the construction of optical interconnects and data communication at the chip scale. Nanowire lasers are built from III–V semiconductor heterostructures. Their unique 1D configuration and high refractive index allow for low optical loss and recirculation in the active nanowire core region. This enables subwavelength laser sizes of only a few hundred nanometers. Nanowires are Fabry–Perot resonator cavities defined by the end facets of the wire, therefore they do not require polishing or cleaving for high-reflectivity facets as in conventional lasers.

Nanowire lasers are nano-scaled lasers with potential as optical interconnects and optical data communication on chip. As stated above, nanowire lasers are built from III–V semiconductor heterostructures. The high refractive index allows for low optical loss in the nanowire core. Nanowire lasers are subwavelength lasers of only a few hundred nanometers. Recent developments have demonstrated repetition rates greater than 200 GHz offering possibilities for optical chip level communications.

Nanowire lasers are coherent light sources (single mode optical waveguides) as any other laser device, with the advantage of operating at the nanoscale. Built by molecular beam epitaxy, nanowire lasers offer the possibility for direct integration on silicon, and the construction of optical interconnects and data communication at the subatomic level.

Recommended videos:

https://www.bing.com/videos/search?q=Nanowire+Lasers&&view=detail&mid=1386C71A25439385B1381386C71A25439385B138&&FORM=VRDGAR

Sensing of proteins and chemicals using semiconductor nanowires:

In an analogous way to FET devices in which the modulation of conductance (flow of electrons/holes) in the semiconductor, between the input (source) and the output (drain) terminals, is controlled by electrostatic potential variation (gate-electrode) of the charge carriers in the device conduction channel, the methodology of a Bio/Chem-FET is based on the detection of the local change in charge density, or so-called "field effect", that characterizes the recognition event between a target molecule and the surface receptor.

This change in the surface potential influences the Chem-FET device exactly as a "gate" voltage does, leading to a detectable and measurable change in the device conduction. When these devices are fabricated using semiconductor nanowires as the transistor element the binding of a chemical or biological species to the surface of the sensor can lead to the depletion or accumulation of charge carriers in the "bulk" of the nanometer diameter nanowire i.e. (small cross section available for conduction channels). Moreover, the wire, which serves as a tunable conducting channel, is in close contact with the sensing environment of the target, leading to a short response time, along with orders of magnitude increase in the sensitivity of the device as a result of the huge S/V ratio of the nanowires.

While several inorganic semiconducting materials such as Si, Ge, and metal oxides (e.g. In2O3, SnO2, ZnO, etc.) have been employed for the preparation of nanowires, Si is usually the material of choice when fabricating nanowire FET-based chemo/biosensors. In short, we can "tune" nanowires to detect different things.

Several examples of the use of silicon nanowire (SiNW) sensing devices include the ultra-sensitive, real-time sensing of biomarker proteins for cancer, detection of single virus particles, and the detection of nitro-aromatic explosive materials such as 2,4,6 Tri-nitrotoluene (TNT) in sensitives superior to these of canines. Silicon nanowires could also be used in their twisted form, as electromechanical devices, to measure intermolecular forces with great precision. Can it sense more than only one subject? The outlook is quite promising, and a number of technologies and devices have been developed already.

Think of medicine, military, security and so many other fields.

Recommended videos:

https://www.bing.com/videos/search?q=Sensing+of+proteins+and+chemicals+us ing+semiconductor+nanowires&&view=detail&mid=7F0EE1469A583A90B3C3 7F0EE1469A583A90B3C3&&FORM=VRDGAR

https://www.bing.com/videos/search?q=Sensing+of+proteins+and+chemicals+us ing+semiconductor+nanowires&&view=detail&mid=833D69B8258316DC2DE8 833D69B8258316DC2DE8&&FORM=VRDGAR

Semiconductor Nanowires:

Semiconductor nanowires (NWs) have exceptional electronic and optical properties, which make them ideal candidates for next-generation nano-electric and photonic devices. Even though those properties are exceptional already, they could be induced even further. When these devices are fabricated using **semiconductor nanowires** as the transistor element the binding of a chemical or biological species to the surface of the sensor can lead to the depletion or accumulation of charge carriers in the "bulk" of the **nanometer** diameter **nanowire** i.e. (small cross section available for conduction channels).

Semiconductor nanowires are high aspect ratio wire-like structures that are often referred to as one-dimensional (1D) materials. The small diameter and cylindrical geometry of **nanowires** can be exploited to develop unique device structures such as axial and radial heterojunctions where the composition and doping are modulated along the length and/or across the radius of the **nanowire**.

Semiconductor nanowires and nanotubes exhibit novel electronic and optical properties owing to their unique structural one-dimensionality and possible quantum confinement effects in two dimensions. With a broad selection of compositions and band structures, these one-dimensional semiconductor nanostructures are the critical components in a wide range of potential nanoscale device applications. To fully exploit these one-dimensional nanostructures, current research has focused on rational synthetic control of one-dimensional nanoscale building blocks, novel properties characterization and device fabrication based on nanowire building blocks, and integration of nanowire elements into complex functional architectures. Significant progress has been made in a few short years.

This review highlights the recent advances in the field, using work from this laboratory for illustration. The understanding of general nanocrystal growth mechanisms serves as the foundation for the rational synthesis of semiconductor heterostructures in one dimension. Availability of these high-quality semiconductor nanostructures allows systematic structural-property correlation investigations, particularly of a size- and dimensionality-controlled nature. Novel properties including nanowire microcavity lasing, phonon transport, interfacial stability and chemical sensing are surveyed.

Some **nanowires** are very good conductors or **semiconductors**, and their miniscule size means that manufacturers could fit millions more transistors on a single microprocessor. As a result, computer speed would increase dramatically. **Nanowires** may play an important role in the field of quantum computers.

Semiconductor Nanowires for Optoelectronics **Applications** have played an important role in the development of information and communications technology, solar cells, solid state lighting and so many other scientific, industrial, and consumer applications.

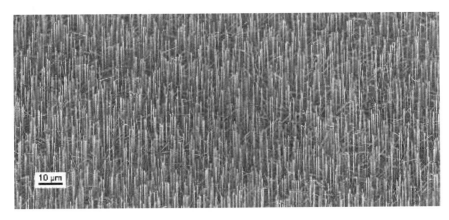

NANOWIRES

Recommended videos:

https://www.bing.com/videos/search?q=Semiconductor+Nanowires&&view=detail&mid=9AAA691D00ADE482440E9AAA691D00ADE482440E&&FORM=VRDGAR

https://www.bing.com/videos/search?q=Semiconductor+Nanowires&&view=detail&mid=06BAF9069FFEBE9F42FE06BAF9069FFEBE9F42FE&&FORM=VDRVRV

https://www.bing.com/videos/search?q=Semiconductor+Nanowires&&view=detail&mid=192A28A78A8999691478192A28A78A8999691478&&FORM=VRDGAR

Band Gap Structure of Semiconductor NWs:

A semiconductor is a material with a small but non-zero band gap that behaves as an insulator at absolute zero but allows thermal excitation of electrons into its conduction band at temperatures that are below its melting point. The melting point would change with different materials and that is another tool that could be used for developing some new applications for semiconductors. In contrast, a material with a large band gap is an insulator. And, it is very difficult or even impossible to change that state applying different temperatures.

In solid-state physics, a band gap, also called an energy gap or bandgap, is an energy range in a solid where no electron states can exist. In graphs of the electronic band structure of solids (example bellow), the band gap generally refers to the energy difference (in electron volts) between the top of the valence band and the bottom of the conduction band in insulators and semiconductors. The picture bellow clearly illustrates that status. The insulators have the biggest gap while the conductors have the smallest one.

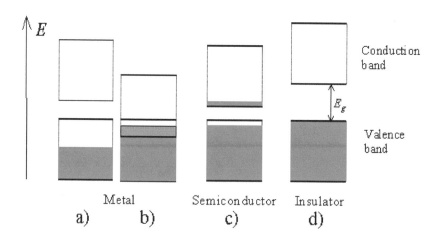

Recommended videos:

https://www.bing.com/videos/search?q=Band+Gap+Structure+of+Semiconducto r+NWs&&view=detail&mid=C617DFCE84E583B9AEDBC617DFCE84E583B 9AEDB&&FORM=VRDGAR

https://www.bing.com/videos/search?q=Band+Gap+Structure+of+Semiconducto r+NWs&&view=detail&mid=FD7352B2ED82E63AFED4FD7352B2ED82E63A FED4&&FORM=VRDGAR

https://www.bing.com/videos/search?q=Band+Gap+Structure+of+Semiconducto r+NWs&&view=detail&mid=2D7C7E7A635A49CF1B5F2D7C7E7A635A49CF 1B5F&&FORM=VDRVRV

7.0   Nanoscale Characterization

Introduction:

Methods used to visualize and manipulate nanomaterials have been a significant factor in the emergence of nanoscience and nanotechnology. Normal light microscopes are limited in depth and resolution by the fact that they cannot see anything much smaller than the wavelengths of visible light and that is not much by today' standards. The smaller features observed with light, microscopes are more dependent upon the smallest wavelength of visible light utilized by the lenses. That is a huge limiting factor. Resolution and resolving power are two terms used to describe the power of the microscope.

Resolution refers to the smallest distinguishable distance between two objects. The resolving power of a microscope involves the best resolution achieved under optimum conditions. Thus, this property is inherent to the instrument used and the method it employs. There are several forms of microscopy available for the study of nanoscale natter, including specialized forms of optical microscopies, electron microscopies, and scanning probe microscopies. All of these technics are still applicable today.

Recommended videos:

https://www.bing.com/videos/search?q=nanoscale+characterization&&view=detail&mid=F0713D3FFCB6010EC797F0713D3FFCB6010EC797&&FORM=VRDGAR

https://www.bing.com/videos/search?q=nanoscale+characterization&&view=detail&mid=B22594F91738EF064452B22594F91738EF064452&&FORM=VRDGAR

Scanning Tunneling Microscopy:

A scanning tunneling microscope (STM) is an instrument for imaging surfaces at the atomic level. Its development in 1981 earned its inventors, Gerd Binnig and Heinrich Rohrer (at IBM Zürich), the Nobel Prize in Physics in 1986. For an STM, good resolution is considered to be 0.1 nm lateral resolution and 0.01 nm depth resolution. A scanning probe microscope has a **sharp probe tip on the end of a cantilever**, which can scan the surface of the specimen. The tip moves back and forth in a very controlled manner and it is possible to move the probe atom by atom.

Scanning probe microscopy is used to create images of nanoscale surfaces and structures or manipulate atoms to move them in specific patterns. It involves a physical probe that scans over the surface of a specimen gathering data that is employed to generate the image or manipulate the atoms. Scanning probe microscopes work differently than optical microscopes because the operator does not have a direct view of the surface but an image that represents the structure of the surface. They are very powerful and can have a very high resolution, up to a nanometer.

There are several different types of scanning probe microscopes including:

- Atomic force microscope (AFM): measures the electrostatic force between the tip and the specimen.
- Magnetic force microscope (MFM): measures the magnetic force between the tip and the specimen.
- Scanning tunneling microscope (STM): measures the electrical current between the tip and the specimen.

Recommended videos:

https://www.bing.com/videos/search?q=Scanning+Tunneling+Microscopy&&view=detail&mid=F5DD2698033DA24E2C55F5DD2698033DA24E2C55&&FORM=VRDGAR

https://www.bing.com/videos/search?q=Scanning+Tunneling+Microscopy&&view=detail&mid=B1B70C531CC93B1722ACB1B70C531CC93B1722AC&&FORM=VDRVRV

Tunneling:

Tunneling is a quantum mechanical effect. A tunneling current occurs when electrons move through a barrier that they classically shouldn't be able to move through. In classical terms, if you don't have enough energy to move "over" a barrier, you won't. However, in the quantum mechanical world, electrons have wavelike properties. These waves don't end abruptly at a wall or barrier but taper off quickly. If the barrier is thin enough, the probability function may extend into the next region, through the barrier!

Because of the small probability of an electron being on the other side of the barrier, given enough electrons, some will indeed move through and appear on the other side. When an electron moves through the barrier in this fashion, it is called tunneling.

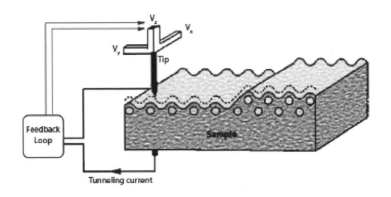

Schematic of scanning tunneling microscopy (STM)

The probe tip, usually made of tungsten (W) or platinum-iridium (Pt-Ir) alloy, is attached to a scanner consisting of three piezoelectric transducers. Each transducer is for x, y, and z tip motion. Upon applying a voltage, a piezoelectric transducer expends or contracts. Applying voltages to the **x** and **y** piezo allows the tip to scan the **x-y** plane. The result is an image of the electron cloud of the surface atoms. STM tips can move atoms around, which is accomplished by varying tip-sample distance and using a voltage bias to pick up and carry, or drag, a surface atom to a new position.

Recommended videos:

https://www.bing.com/videos/search?q=Tunneling&&view=detail&mid=92C43C2F15D434EBD29492C43C2F15D434EBD294&&FORM=VRDGAR

STM operation modes - scanning tunneling microscopy (STM) **has five main variable parameters**. These are horizontal coordinates x and y, height z, the bias voltage V and tunneling current I. If there is an inhomogeneous compound in the surface the work function will be inhomogeneous as well. This alters the local barrier height. By using the two modes described above we would get a virtual hole or adatom.

Feedback Loop:

Feedback occurs when outputs of a system are routed back as inputs as part of a chain of cause-and-effect that forms a circuit or loop. The system can then be said to feed back into itself. **Feedback loop** - a circuit that feeds back some of the output to the input of a system. **feedback** circuit. circuit, electric circuit, electrical circuit - an electrical device that provides a path for electrical current to flow. control circuit, negative **feedback** circuit - a **feedback** circuit that subtracts from the input. Channel or pathway formed by an "effect" returning to its "cause," and generating either more or less of the same effect. A dialogue is an example of a **feedback loop**.

A positive feedback loop occurs in nature when the product of a reaction leads to an increase in that reaction. If we look at a system in homeostasis, a positive feedback loop moves a system further away from the target of equilibrium. It does this by amplifying the effects of a product or event and occurs when something needs to happen quickly.

A negative feedback loop occurs in biology when the product of a reaction leads to a decrease in that reaction. In this way, a negative feedback loop brings a system closer to a target of stability or homeostasis. Negative feedback loops are responsible for the stabilization of a system, and ensure the maintenance of a steady, stable state. The response of the regulating mechanism is opposite to the output of the event.

Recommended videos:

https://www.bing.com/videos/search?q=Feedback+Loop+&&view=detail&mid=0C7FBDBB8B2141A257770C7FBDBB8B2141A25777&&FORM=VRDGAR

https://www.bing.com/videos/search?q=Feedback+Loop+++&&view=detail&mid=8682AB196B73897F0DC28682AB196B73897F0DC2&&FORM=VRDGAR

https://www.bing.com/videos/search?q=Operating+Modes+of+AFM&&view=detail&mid=81ADB90B8367EB0B3DC081ADB90B8367EB0B3DC0&&FORM=VRDGAR

AFM:

AFM is an atomic force microscope with controlling computer and other supporting systems. AFM is a type of scanning probe microscopy (SPM), with demonstrated resolution on the order of fractions of a nanometer, more than 1000 times better than the optical diffraction limit. So, it's a thousand times better than any optical microscope. What a power; what an opportunity for our research ambitions. The information is gathered by "feeling" or "touching" the surface with a mechanical probe. Piezoelectric elements that facilitate tiny but accurate and precise movements on (electronic) command enable precise scanning.

The AFM has three major abilities: force measurement, imaging, and manipulation:

In force measurement, AFMs can be used to measure the forces between the probe and the sample as a function of their mutual separation. This can be applied to perform force spectroscopy, to measure the mechanical properties of the sample, such as the sample's Young's modulus, a measure of stiffness.

For imaging, the reaction of the probe to the forces that the sample imposes on it can be used to form an image of the three-dimensional shape (topography) of a sample surface at a high resolution. This is achieved by raster scanning the position of the sample with respect to the tip and recording the height of the probe that corresponds to a constant probe-sample interaction. The surface topography is commonly displayed as a pseudo-color plot.

Pseudo-color Plots. It is "pseudo" (meaning false) color because the colors used refer to the density of the cells relative to one another, not to the spectra that the cells emit. Blue and green correspond to areas of lower cell density, red and orange are areas of high cell density, and yellow is mid-range.

In manipulation, the forces between tip and sample can also be used to change the properties of the sample in a controlled way. Yes, we can move particles around to change the basic structure and to create something more helpful to our goals. Yes, we can set the goals and achieve them. That often was not possible before. Some examples of this include atomic manipulation, scanning probe lithography and local stimulation of cells.

Simultaneous with the acquisition of topographical images, other properties of the sample can be measured locally and displayed as an image, often with similarly high resolution. Examples of such properties are mechanical properties like stiffness or adhesion strength and electrical properties such as conductivity or surface potential. In fact, the majority of SPM techniques are extensions of AFM that use this modality.

Recommended videos:

https://www.bing.com/videos/search?q=nanoscale+characterization&&view=detail&mid=5AF0D317083C70D46F3E5AF0D317083C70D46F3E&&FORM=VRDGAR

https://www.bing.com/videos/search?q=scanning+tunneling+Feedback+Loop+&&view=detail&mid=60245AF295F4CC71A55160245AF295F4CC71A551&&FORM=VRDGAR

AFM Components:

AFMs consist of a piezoelectric scanner moving a tip across the surface of a sample. A detector is used to monitor tip-sample interaction. A control station, which includes a computer and an AFM controller-electronics unit, controls the AFM's operation and generates digital images.

Main Components of an AFM

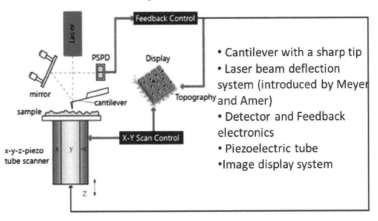

- Cantilever with a sharp tip
- Laser beam deflection system (introduced by Meyer and Amer)
- Detector and Feedback electronics
- Piezoelectric tube
- Image display system

Recommended videos:

https://www.bing.com/videos/search?q=AFM+Components+&&view=detail&mid=91D352A329636DF877FA91D352A329636DF877FA&&FORM=VRDGAR

https://www.bing.com/videos/search?q=AFM+Components+&&view=detail&mid=2818E0C00A93996403E12818E0C00A93996403E1&&FORM=VRDGAR

AFM Operation:

Atomic force microscopy or scanning force microscopy is a very-high-resolution type of scanning probe microscopy, with demonstrated resolution on the order of fractions of a nanometer, more than 1000 times better than the optical diffraction limit. The AFM consists of a cantilever with a sharp tip at its end that is used to scan the specimen surface, and it happens very fast. The cantilever is typically silicon or silicon nitride with a tip radius of curvature on the order of nanometers. When the tip is brought into proximity of a sample surface, forces between the tip and the sample lead to a deflection of the cantilever according to Hooke's law.

Hooke's law is a principle of physics that states that the force F needed to extend or compress a spring by some distance X is proportional to that distance. That is: F = kX, where k is a constant factor characteristic of the spring: its stiffness, and X is small compared to the total possible deformation of the spring. The law is named after 17th-century British physicist Robert Hooke.

The AFM principle is based on the cantilever/tip assembly that interacts with the sample. This assembly is also commonly referred to as the probe and has quite a specific configuration. The AFM probe interacts with the substrate through a raster scanning motion. The up/down and side to side motion of the AFM tip as it scans along the surface is monitored through a laser beam reflected off the cantilever. This reflected laser beam is tracked by a position sensitive photo-detector (PSPD) that picks up the vertical and lateral motion of the probe. The deflection sensitivity of these detectors must be calibrated in terms of how many nanometers of motion correspond to a unit of voltage measured on the detector. Computers control all that.

In order to achieve the AFM modes known as **tapping modes**, the probe is mounted into a holder with a shaker piezo. The shaker piezo provides the ability to oscillate the probe at a wide range of frequencies (typically 100 Hz to 2 MHz). Tapping modes of operation can be divided into **resonant modes** (where operation is at or near the resonance frequency of the cantilever) and **off-resonance modes** (where operation is at a frequency usually far below the cantilever's resonance frequency). Both frequencies are employed equally but in different situations.

The principle of how AFM works is depicted in the following schematic:

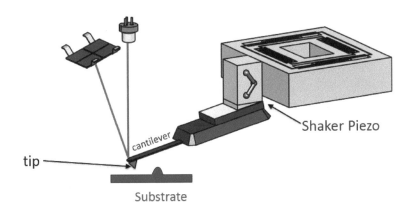

tip

cantilever

Shaker Piezo

Substrate

Recommended videos:

https://www.bing.com/videos/search?q=AFM+Operation&&view=detail&mid=6
0245AF295F4CC71A55160245AF295F4CC71A551&&FORM=VDRVRV

https://www.bing.com/videos/search?q=AFM+Operation&&view=detail&mid=C
D1D60431D1E3B694FDECD1D60431D1E3B694FDE&rvsmid=60245AF295F
4CC71A55160245AF295F4CC71A551&FORM=VDRVRV

https://www.bing.com/videos/search?q=AFM+Operation&&view=detail&mid=D
8B5EE80D37F5B7559E6D8B5EE80D37F5B7559E6&&FORM=VDRVRV

Operating Modes of AFM:

The most commonly used **modes** of **operation** of an **AFM** are: Contact **mode AFM** (aka C-**AFM** or CMAFM) Tapping **mode AFM** (aka TMAFM, IC-**AFM** or AM-**AFM**) Noncontact **mode** (aka NC-**AFM**, close contact **AFM** or FM-**AFM**).

When the oscillating probe approaches the sample surface, the oscillation changes due to the interaction between the probe and the forces from the sample. This leads to reduction in the frequency and amplitude of the oscillation. The oscillation is monitored by the optical sensor, and the scanner adjusts the **z** height via feedback loop to maintain a constant vibration amplitude.

Force curves (force-versus-distance curve) typically show the **deflection of the free end of the AFM cantilever** as the fixed end of the cantilever is brought vertically towards and then away from the sample surface. Force curve data allows scientists to study nanomechanical phenomena. In these experiments, one end of the molecule is attached to the AFM tip and the other end is attached to the surface of a substrate. Data collected from force curve measurements includes adhesive forces and rapture forces needed to break the bonds in a molecule.

Force – Distance Graph

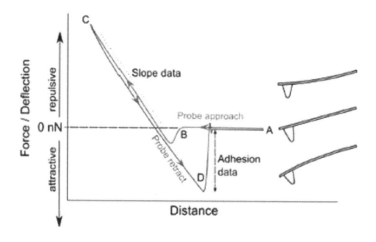

A- The cantilever is lowered to the surface;

B- Tip snaps into contact when it is 1-10 nm away from the sample surface;

C- As the cantilever is lowered, the cantilever bends up because the tip is pushing against the sample surface;

D- The cantilever is pulled away from the sample surface and the tip holds onto the surface causing the cantilever to bend down due to intermolecular interactions between the surface atoms and tip atoms;

The sharpness of tips employed in atomic force microscopes has allowed researches to move atoms on surfaces and fabricate nanopatterns in thin films. In nano-shaving, the AFM tip exerts high force on a sample surface. This pressure causes the displacement of absorbed molecules or thin films. Thus, holes and trenches can be fabricated. This procedure allows precise control over the size and geometry of the fabricated features. In addition, this technique makes it possible to obtain an edge resolution better than 2 nm.

Recommended videos:

https://www.bing.com/videos/search?q=Modes+of+AFM&&view=detail&mid=0 7BDBA04A12C05BB8CDC07BDBA04A12C05BB8CDC&&FORM=VRDGAR

https://www.bing.com/videos/search?q=afm+modes&&view=detail&mid=92E15 D894A6D6F3C4CA792E15D894A6D6F3C4CA7&&FORM=VRDGAR

https://www.bing.com/videos/search?q=afm+modes&&view=detail&mid=83525 9548BECC9EA9CB4835259548BECC9EA9CB4&&FORM=VRDGAR

https://www.bing.com/videos/search?q=afm+modes&&view=detail&mid=18CD0 AD6F7EC0A888DC418CD0AD6F7EC0A888DC4&&FORM=VRDGAR

https://www.bing.com/videos/search?q=afm+modes&&view=detail&mid=E405C F7FD69136587FE5E405CF7FD69136587FE5&&FORM=VRDGAR

https://www.bing.com/videos/search?q=afm+modes&&view=detail&mid=EB3B 15D46635E7D29C27EB3B15D46635E7D29C27&&FORM=VRDGAR

Scanning Electron Microscopy:

A scanning electron microscope (SEM) is a type of electron microscope that produces images of a sample by scanning it with a focused beam of electrons. That is a very thin beam of electrons that comfortably operates on nano scale. The electrons of the beam interact with atoms in the sample, producing various signals that contain information about the sample's surface topography and composition.

The original form of the electron microscope, the **transmission electron microscope (TEM)**, uses a high voltage electron beam to illuminate the specimen and create an image. The electron beam is produced by an electron gun, commonly fitted with a tungsten filament cathode as the electron source. The higher the power of that beam, the better resulting picture. Thus, we need to create that power and sustain it for long enough to produce the necessary pictures. **Electron gun.** Electron guns may be classified by the type of electric field generation (DC or RF), by emission mechanism (thermionic, photocathode, cold emission, plasmas source), by focusing (pure electrostatic or with magnetic fields), or by the number of electrodes.

Scanning electron microscopes (SEMs) employ a lower-energy electron beam, but it can still be damaging to the sample. The vacuum inside an electron microscope is important for its function and the clarity of operation. Without a vacuum, there would be particles of other materials present, thus electrons being aimed at the sample would be deflected (knocked off course) when they hit the other particles in the tube. It will result in a non-clear and often incorrect picture.

Thus, operation in vacuum is a must and the Electron Microscopes are made with that in mind. An electron microscope is designed for the small-scale recognition, so precise and allows us to see at these **small-scales** with such a clear definition. Electron Microscopes work by using an electron beam instead of visible light and an electron detector instead of our eyes. The visible light tends to disburse with no concentration while the electron beam is focused. An electron beam allows us to see at very small scales because electrons can also behave as light.

Electron microscopes are **used** to investigate the ultrastructure of a wide range of biological and inorganic specimens including microorganisms, cells, large molecules, biopsy samples, metals, and crystals. Industrially, **electron microscopes** are often **used** for quality control and failure analysis.

How to Use a Scanning Electron Microscope – Steps:

- Obtain a prepared sample;
- Bring the SEM to atmospheric pressure in order to open the sample door;
- Put on gloves;
- While holding the door shut, press and hold the EVAC button until it begins to blink;
- After the 5 minutes is up, you are ready to turn on the electron beam and start taking pictures!

The scanning electron microscope (SEM) uses a focused beam of high-energy electrons to generate a variety of signals at the surface of solid specimens. The signals that derive from electron-sample interactions reveal information about the sample including external morphology (texture), chemical composition and crystalline structure, and orientation of materials making up the sample. In most applications, data are collected over a selected area of the surface of the sample, and a 2-dimensional image is generated that displays spatial variations in these properties. If the sample is too large for one picture, the scanner could be rolled over the sample or smaller pictures could be fitted together. This approach is presently used even in archeology to outline the discovery.

Areas ranging from approximately 1 cm to 5 microns in width can be imaged in a scanning mode using conventional SEM techniques (magnification ranging from 20X to approximately 30,000X, spatial resolution of 50 to 100 nm). The SEM is also capable of performing analyses of selected point locations on the sample; this approach is especially useful in qualitatively or semi-quantitatively determining chemical compositions (using EDS), crystalline structure, and crystal orientations (using EBSD). The design and function of the SEM is very similar to the EPMA and considerable overlap in capabilities exists between the two instruments.

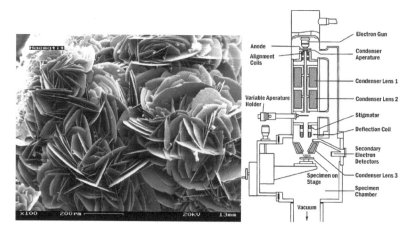

Scanning Electron Microscope and a picture resulting from its work.

Recommended videos:

https://www.bing.com/videos/search?q=Scanning+Electron+Microscopy&&view=detail&mid=F7466D4D8DDFA0B91D90F7466D4D8DDFA0B91D90&&FORM=VRDGAR

https://www.bing.com/videos/search?q=Scanning+Electron+Microscopy&&view=detail&mid=BEF8B9E4DCA9A0D8968EBEF8B9E4DCA9A0D8968E&&FORM=VDRVRV

Electron Gun:

An electron gun (also called electron emitter) is an electrical component in some vacuum tubes that produces a narrow, collimated electron beam that has a precise kinetic energy. The largest use is in cathode ray tubes (CRTs), used in nearly all television sets, computer displays and oscilloscopes that are not flat-panel displays.

Electron gun is defined as the source of focused and accelerated **electron** beam. It is a device used in Cathode Ray Tube for displaying the image on the phosphorous screen of CRT. The **electron gun** emits electrons and forms them into a beam by the help of a heater, cathode, grid, pre-accelerating, accelerating and focusing anode.

X-Rays:

X-Ray is an electromagnetic wave of high energy and very short wavelength, which is able to pass through many materials opaque to light. X-radiation (composed of X-rays) is a form of electromagnetic radiation. Most X-rays have a wavelength ranging from 0.01 to 10 nanometers, corresponding to frequencies in the range 30 petahertz to 30 exahertz and energies in the range 100 eV to 100 keV. X-ray wavelengths are shorter than those of UV rays and typically longer than those of gamma rays. X-rays are produced when electrons strike a metal target. The electrons are liberated from the heated filament and accelerated by a high voltage towards the metal target. When the electrons strike the target, their energy is converted to X-rays. X-rays can also be produced by a synchrotron, a type of particle accelerator that causes charged particles to move in a closed, circular path. When high-speed electrons are forced to move in a circular path by a magnetic field, they create a beam called X-Ray.

There are many types of X-rays that are used to diagnose conditions and diseases. The following are examples. **Mammography** is a type of X-ray radiograph that is used to detect breast cancer; **Computed Tomography** (**CT**) scans combine X-ray with computer processing to create detailed pictures (scans) of cross sections of the body that are combined to form a three-dimensional X-ray image. There are also **Fluoroscopy** uses X-rays and a fluorescent screen to study moving or real-time structures in the body, such as viewing the heart beating. It can also be used in combination with swallowed or injected contrast agents to view the digestive processes or blood flow. Cardiac angioplasty uses fluoroscopy with a contrast agent to guide an internally threaded catheter to help open clogged arteries. Fluoroscopy is also used to precisely place instruments in certain locations within the body, such as during epidural injections or joint aspirations.

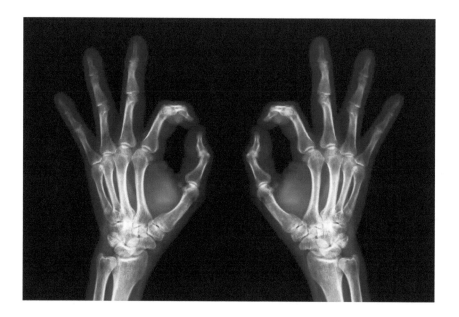

Other uses for X-rays and other types of radiation include cancer treatment. High-energy radiation in much higher doses than what is used for X-ray imaging may be utilized to help destroy cancerous cells and tumors by damaging their DNA.

Radiation does have some risks to consider, but it is also important to remember that X-rays can help detect disease or injury at early stages so the ailment can be treated timely and appropriately. Sometimes X-ray testing can be life-saving. The risk from X-rays comes from the radiation they produce, which can harm living tissues. This risk is relatively small, if X-Ray is employed properly, but it increases with cumulative exposure. That is, the more you are exposed to radiation over your lifetime, the higher your risk of harm from the radiation.

Recommended videos:

https://www.bing.com/videos/search?q=X-Rays&&view=detail&mid=C6568D062C5C4DFDE9C2C6568D062C5C4DFDE9C2&&FORM=VRDGAR

https://www.bing.com/videos/search?q=X-Raysview=detail&mid=C65633CB05EA48691CCDC65633CB05EA48691CCD&&FORM=VDRVRV

https://www.bing.com/videos/search?q=X-Rays&&view=detail&mid=17DB8150E30A9BC1612417DB8150E30A9BC16124&rvsmid=B926DEF0DEAE843BCCD1B926DEF0DEAE843BCCD1&FORM=VDRVRV

https://www.bing.com/videos/search?q=X-Rays&&view=detail&mid=49B94C0172AF9F19EE4149B94C0172AF9F19EE41&&FORM=VDRVRV

TEM:

Transmission electron microscopy (TEM) is a microscopy technique in which a beam of electrons is transmitted through an ultra-thin specimen, interacting with the specimen as it passes through it. It could be considered a form of X-Ray and often employed for the same purpose.

Transmission electron microscopy (**TEM**, an abbreviation which can also stand for the instrument, a **transmission electron microscope**) is a microscopy technique in which a beam of electrons is transmitted through a specimen to form an image. The specimen is most often an ultrathin section less than 100 nm thick or a suspension on a grid.

An image is formed from the interaction of the electrons with the sample as the beam is transmitted through the specimen. The image is then magnified and focused onto an imaging device, such as a fluorescent screen, a layer of photographic film, or a sensor such as a scintillator attached to a charge-coupled device.

Transmission electron microscopes are capable of imaging at a significantly higher resolution than light microscopes, owing to the smaller de Broglie wavelength of electrons. This enables the instrument to capture fine detail - even as small as a single column of atoms, which is thousands of times smaller than a resolvable object seen in a light microscope. Light microscopes have the relatively low limits. Transmission electron microscopy is a major analytical method in the physical, chemical and biological sciences. TEMs find application in cancer research, virology, and materials science as well as pollution, nanotechnology and semiconductor research, but also in other fields such as paleontology and palynology (the study of pollen grains and other spores, especially as found in archaeological or geological deposits).

TEM instruments boast an enormous array of operating modes including conventional imaging, scanning TEM imaging (STEM), diffraction, spectroscopy, and combinations of these. Even within conventional imaging, there are many fundamentally different ways that contrast is produced, called "image contrast mechanisms". Contrast can arise from position-to-position differences in the thickness or density ("mass-thickness contrast"), atomic number ("Z contrast", referring to the common abbreviation Z for atomic number), crystal structure or orientation ("crystallographic contrast" or "diffraction contrast").

The slight quantum-mechanical phase shifts that individual atoms produce in electrons that pass through them ("phase contrast"). The energy lost by electrons on passing through the sample ("spectrum imaging") and more. Each mechanism tells the user a different kind of information, depending not only on the contrast mechanism but on how the microscope is employed -.the settings of lenses, apertures, and detectors. What this means is that a TEM can return an extraordinary variety of nanometer - and atomic-resolution information, in ideal cases revealing not only where all the atoms are but what kinds of atoms they are and how they are bonded to each other. For this reason, TEM is regarded as an essential tool for nanoscience in both biological and materials fields. Still, we are limited to what we have; this field is booming with new developments.

Recommended videos:

*https://www.bing.com/videos/search?q=TEM&&view=detail&mid=9B90BFDFC
F6C69C0ADF89B90BFDFCF6C69C0ADF8&&FORM=VRDGAR*

*https://www.bing.com/videos/search?q=Tem+Microscope&&view=detail&mid=
E7AB967327E0E7BE9217E7AB967327E0E7BE9217&&FORM=VRDGAR*

TEM Components

A TEM is composed of several components, which include a vacuum system in which the electrons travel, an electron emission source for generation of the electron stream, a series of electromagnetic lenses, as well as electrostatic plates.

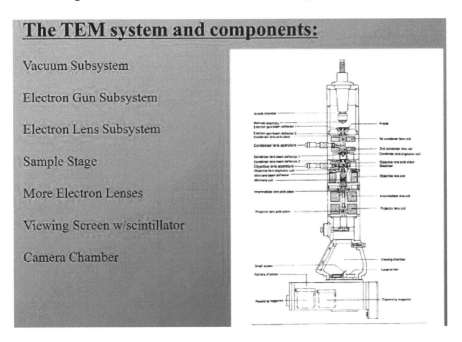

Transmission Electron Microscope (TEM): It consist of a system of electromagnetic lenses mounted in a column. Components of Electron Microscope: EM is in the form of a tall column which is vertically mounted. Generally, it consists of the following main components:

1. Electron gun.

2. Electromagnetic lenses—three sets.

3. Image viewing and recording system.

Recommended videos:

https://www.bing.com/videos/search?q=Transmission+electron+microscopy+Components&&view=detail&mid=FAC8C169710ADD42E7EFFAC8C169710ADD42E7EF&&FORM=VRDGAR

https://www.bing.com/videos/search?q=Transmission+electron+microscopy+Components&&view=detail&mid=27C5986FD30F3789FA1727C5986FD30F3789FA17&&FORM=VDRVRV

8.0   Nanofabrication Techniques

Introduction:

There are two very different paths used in fabrication of nanomaterials. One is a top-down strategy of miniaturizing current technologies, while the other is a bottom-up strategy of building ever-more-complex molecular devices atom by atom. Top-down approaches are good for producing structures with long-range order and for making macroscopic connections, while bottom-up approaches are best suited for assembly and establishing short-range order at nanoscale dimensions. The integration of top-down and bottom-up techniques is expected to eventually provide the best combination of tools for nanofabrication. Nanotechnology requires new tools for fabrication and measurement.

The most common **top-down approach** to fabrication involves lithographic patterning techniques using short-wavelength optical sources. A key advantage of the top-down approach - as developed in the fabrication of integrated circuits - is that the parts are both patterned and built in place, so that no assembly step is needed. Optical lithography is a relatively mature field because of the high degree of refinement in microelectronic chip manufacturing, with current short-wavelength optical lithography techniques reaching dimensions just below 100 nanometers (the traditional threshold definition of the nanoscale). Shorter-wavelength sources, such as extreme ultraviolet and X-ray, are being developed to allow lithographic printing techniques to reach dimensions from 10 to 100 nanometers. Scanning beam techniques such as electron-beam lithography provides patterns down to about 20 nanometers. Here the pattern is written by sweeping a finely focused electron beam across the surface. Focused ion beams are also used for direct processing and patterning of wafers, although with somewhat less resolution than in electron-beam lithography. Still-smaller features are obtained by using scanning probes to deposit or remove thin layers.

Bottom-up, or self-assembly, approaches to nanofabrication use chemical or physical forces operating at the nanoscale to assemble basic units into larger structures. As component size decreases in nanofabrication, bottom-up approaches provide an increasingly important complement to top-down techniques. Inspiration for bottom-up approaches comes from biological systems, where nature has harnessed chemical forces to create essentially all the structures needed by life. Researchers hope to replicate nature's ability to produce small clusters of specific atoms, which can then self-assemble into more-elaborate structures.

Recommended videos:

https://slideplayer.com/slide/5850112/

https://www.bing.com/videos/search?q=Nanofabrication+Techniques&&view=detail&mid=6503F2131CE3361FBA976503F2131CE3361FBA97&&FORM=VRDGAR

https://www.bing.com/videos/search?q=Nanofabrication+Techniques&&view=detail&mid=BF278C32F198E7995DE6BF278C32F198E7995DE6&&FORM=VRDGAR

https://www.bing.com/videos/search?q=Nanofabrication+Techniques&&view=detail&mid=96807463EA569CCEEB2496807463EA569CCEEB24&&FORM=VRDGAR

https://www.bing.com/videos/search?q=Photolithography&&view=detail&mid=F0027E3977EAD050513DF0027E3977EAD050513D&&FORM=VRDGAR

Soft Lithography:

In technology, soft lithography refers to a family of techniques for fabricating or replicating structures using "elastomeric stamps, molds, and conformable photomasks". It is called "soft" because it uses elastomeric materials, most notably PDMS.

Soft lithography, however, extends the possibilities of conventional photolithography. Unlike **photolithography**, **soft lithography** can process a wide range of elastomeric materials, i.e. mechanically soft materials. Therefore, the term "soft" is used. For instance, **soft lithography** is well suited for polymers, gels, and organic monolayers.

Recommended videos:

https://www.bing.com/videos/search?q=Nanofabrication+Techniques&&view=detail&mid=D089485F26A3EFD77734D089485F26A3EFD77734&&FORM=VRDGAR

https://www.bing.com/videos/search?q=Photolithography&&view=detail&mid=D2772ED8503378BF2AE1D2772ED8503378BF2AE1&&FORM=VRDGAR

https://www.bing.com/videos/search?q=Nanolithography&&view=detail&mid=F2BF07D79875A71C485BF2BF07D79875A71C485B&&FORM=VRDGAR

https://www.bing.com/videos/search?q=Nanolithography&&view=detail&mid=366DB5EE58964EDC4A76366DB5EE58964EDC4A76&&FORM=VRDGAR

Microcontact Printing:

Microcontact printing (or μCP) is a form of soft lithography that uses the relief patterns on a master polydimethylsiloxane (PDMS) stamp to form patterns of self-assembled monolayers (SAMs) of ink on the surface of a substrate through conformal contact as in the case of nano-transfer printing (nTP). Its applications are wide ranging including microelectronics, surface chemistry and cell biology.

Microcontact printing is a lithographic technique for the fabrication of a variety of microelectronic components, such as electrodes for organic field effect transistors. The similar process is often used for much larger applications that are employed in different industries as an on-line printing process.

In addition, stamps for microcontact printing can be rapidly produced by injection molding using a poly(urethane) resin, with a silicon master as the mold insert. A snapshot of the multi-part injection mold used in this study is shown in. Depending on the needs, this process could be much more complex.

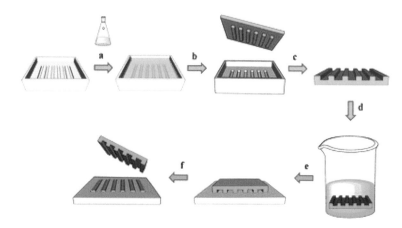

Microcontact printing is a type of soft lithography and replica molding procedure. In this process, an ink solution is transferred from an elastomeric mold, or stamp, to a substrate surface. The stamp is produced using soft lithography approaches and the ink solution is typically composed of proteins, protein mixtures or small molecules.

Recommended videos:

https://www.bing.com/videos/search?q=Micro+Contact+Printing&&view=detail&mid=66
6E852E8BA19997704E666E852E8BA19997704E&&FORM=VRDGAR

https://www.bing.com/videos/search?q=Microcontact+Printing&&view=detail&mid=216
289D4DE292A9FF87A216289D4DE292A9FF87A&&FORM=VRDGAR

https://www.bing.com/videos/search?q=Microcontact+Printing&&view=detail&mid=1E9
BCEFF821FCCEA36D41E9BCEFF821FCCEA36D4&&FORM=VDRVRV

Micromolding in Capillaries (MIMIC):

Micromolding in capillaries (MIMIC) is a novel, simple and convenient procedure for crystallizing microspheres from latex suspensions onto a support, and for delivering microspheres to and assembling them in geometrically confined regions on the surface of substrate.

A **fluidal** precursor to a polymer, a solution, or a suspension of the material to be patterned filling these channels by **capillary** action. After the material in the fluid had cross-linked, crystallized, cured, adhered, or deposited onto the surface of the substrate, the elastomeric component was removed.

Physical Vapor Deposition:

Physical vapor deposition (PVD) describes a variety of vacuum deposition methods which can be used to produce thin films. PVD uses physical process (such as heating or sputtering) to produce a vapor of material, which is then deposited on the object which requires coating. PVD is used in the manufacture of items which require thin films for mechanical, optical, chemical or electronic applications. PVD Coating refers to a variety of thin film deposition techniques where a solid material is vaporized in a vacuum environment and deposited on substrates as a pure material or alloy composition coating.

As the process transfers the coating material as a single atom or on the molecular level, it can provide extremely pure and high-performance coatings which for many applications can be preferable to other methods used. At the heart of every microchip, and semiconductor device, durable protective film, optical lens, solar panel and many medical devices, PVD Coatings provide crucial performance attributes for the final product. Whether the coating needs to be extremely thin, pure, durable or clean, PVD provides the solution.

It is used in a wide variety of industries like optical applications ranging from eye glasses to self-cleaning tinted windows, photovoltaic applications for solar energy, device applications like computer chips, displays and communications as well as functional or decorative finishes, from durable hard protective films to brilliant gold, platinum or chrome plating.

The two most common Physical Vapor Deposition Coating processes are Sputtering and Thermal Evaporation. Sputtering involves the bombardment of the coating material known as the target with a high energy electrical charge causing it to "sputter" off atoms or molecules that are deposited on a substrate like a silicon wafer or solar panel. Thermal Evaporation involves elevating a coating material to the boiling point in a high vacuum environment causing a vapor stream to rise in the vacuum chamber and then condense on the substrate.

Recommended videos:

https://www.bing.com/videos/search?q=Physical+Vapor+Deposition&&view=detail&mid=FA8F6B3EA0B5BA9FB05FFA8F6B3EA0B5BA9FB05F&&FORM=VRDGAR

https://www.bing.com/videos/search?q=Physical+Vapor+Deposition&&view=detail&mid=C25EF8871DF13D0BC6EFC25EF8871DF13D0BC6EF&&FORM=VRDGAR

Chemical Vapor Deposition:

Chemical vapor deposition (CVD) is a chemical process used to produce high quality, high-performance, solid materials. The process is often used in the semiconductor industry to produce thin films. In typical CVD, the wafer (substrate) is exposed to one or more volatile precursors, which react and/or decompose on the substrate surface to produce the desired deposit.

Chemical vapor deposition (CVD) is parent to a family of processes whereby a solid material is deposited from a vapor by a chemical reaction occurring on or in the vicinity of a normally heated substrate surface. The resulting solid material is in the form of a thin film, powder, or single crystal.

CVD covers processes such as:

- Atmospheric Pressure Chemical Vapor Deposition (APCVD);

- Low Pressure Chemical Vapor Deposition (LPCVD);

- Metal-Organic Chemical Vapor Deposition (MOCVD);

- Plasma Assisted Chemical Vapor Deposition (PACVD) or Plasma Enhanced Chemical Vapor Deposition (PECVD);

- Laser Chemical Vapor Deposition (LCVD);

- Photochemical Vapor Deposition (PCVD);

- Chemical Vapor Infiltration (CVI);

- Chemical Beam Epitaxy (CBE).

Recommended videos:

https://www.bing.com/videos/search?q=Chemical+Vapor+Deposition+Process&&view=detail&mid=5FE7E1727D6B3A4370B45FE7E1727D6B3A4370B4&&FORM=VRDGAR

https://www.bing.com/videos/search?q=Chemical+Vapor+Deposition+Process&&view=detail&mid=C2A55F29E1C04FADB2A8C2A55F29E1C04FADB2A8&rvsmid=5FE7E1727D6B3A4370B45FE7E1727D6B3A4370B4&FORM=VDRVRV

https://www.bing.com/videos/search?q=Chemical+Vapor+Deposition+Process&&view=detail&mid=7273FC201033CB3D32D17273FC201033CB3D32D1&&FORM=VDRVRV

Etch:

Etch definition is - to produce (something, such as a pattern or design) on a hard material by eating into the material's surface (as by acid or laser beam). To cut into the surface of (glass, for example) by the action of acid, especially by coating the surface with wax or another protective layer and drawing lines with a needle and then using the acid to form the lines on the unprotected parts of the surface. Etching is a process which creates topographical surface features by selectively removing material through physical or chemical means. Etching mechanism utilize liquid or gas-based processes. Etching is isotropic (having a physical property which has the same value when measured in different directions) if it occurs uniformly in all directions.

Etching can be dry and wet. Dry etching techniques are plasma-based and possess several advantages when compared to wet etching. Dry etching is anisotropic and possesses vertical sidewalls. Wet and dry etching involves the transfer of mask patterns onto wafers. Masks ensure that only uncovered areas are etched away, while leaving areas covered by the mask intact.

Recommended videos:

https://www.bing.com/videos/search?q=Semiconductor+Etching&&view=detail&mid=A2BAB177A56DCA0861D7A2BAB177A56DCA0861D7&&FORM=VRDGAR

https://www.bing.com/videos/search?q=Semiconductor+Etching&&view=detail&mid=D04BAE43D11367B13934D04BAE43D11367B13934&&FORM=VRDGAR

https://www.bing.com/videos/search?q=Semiconductor+Etching&&view=detail&mid=AA7E33A22295FC629B52AA7E33A22295FC629B52&&FORM=VDRVRV

E-beam Lithography:

Electron-beam lithography (often abbreviated as e-beam lithography) is the practice of scanning a focused beam of electrons to draw custom shapes on a surface covered with an electron-sensitive film called a resist ("exposing").

Electron beam lithography (e-beam lithography) is a direct writing technique that uses an accelerated **beam** of electrons to pattern features down to **sub-10 nm** on substrates that have been coated with an **electron beam** sensitive resist. Electron beam lithography (**e-beam lithography** or **EBL**) is a versatile tool capable of making almost any kind of pattern imaginable within nanotechnology.

E-beam lithography is a serial process just as any other **beam-based writing** techniques (**ion beam** and **laser**), and the sequential nature of the process makes writing very time consuming and impractical for mass production. Therefore, it is widely used for R&D or pilot production and photomask production for **optical lithography**.

Recommended videos:

https://www.bing.com/videos/search?q=E-beam+Lithography&&view=detail&mid=CE466895884A7E83654CCE46689588 4A7E83654C&&FORM=VRDGAR

https://www.bing.com/videos/search?q=E-beam+Lithography&&view=detail&mid=D3E6D5688F6FE204583BD3E6D568 8F6FE204583B&&FORM=VDRVRV

Focused Ion Beam:

Focused ion beam, also known as FIB, is a technique used particularly in the semiconductor industry, materials science and increasingly in the biological field for site-specific analysis, deposition, and ablation of materials. A FIB setup is a scientific instrument that resembles a scanning electron microscope (SEM).

An ion beam is a type of charged particle beam consisting of ions. Ion beams have many uses in electronics manufacturing (principally ion implantation) and other industries. A variety of ion beam sources exist, some derived from the mercury vapor thrusters developed by NASA in the 1960s. In the **Focused Ion Beam** Systems, the primary **beam** also produces secondary electrons (e-). As the primary **beam** raster on the sample surface, the signal from the sputtered **ions** or secondary electrons is collected to form an image. The image is very deep and in good details because at low primary **beam** currents, very little material is sputtered. This is a huge advantage of the Focused Ion Beam.

Recommended videos:

https://www.bing.com/videos/search?q=Focused+Ion+Beam&&view=detail&mid=48EA7 9213AD5106C66AE48EA79213AD5106C66AE&&FORM=VRDGAR

https://www.bing.com/videos/search?q=Focused+Ion+Beam&&view=detail&mid=2DF0 E35FF25860AC75FF2DF0E35FF25860AC75FF&&FORM=VDRVRV

Photolithography:

Photolithography, also termed optical lithography or UV lithography, is a process used in microfabrication to pattern parts of a thin film or the bulk of a substrate. It uses light to transfer a geometric pattern from a photomask to a light-sensitive chemical "photoresist", or simply "resist," on the substrate. It is responsible for both: the decrease in size and the increase in power of computing systems because it allows the incorporation of more components in an integrated circuit. Photolithography is a process that is photographic in nature. It involves the projection of light onto a mask containing a pattern of an electronic circuit. Once light passes through the mask, the pattern is projected onto a wafer covered with a light-sensitive photoresist.

Recommended videos:

https://www.bing.com/videos/search?q=Photolithography&&view=detail&mid=3
EB5DE52055C55C171153EB5DE52055C55C17115&&FORM=VRDGAR

https://www.bing.com/videos/search?q=Photolithography&&view=detail&mid=1
E734031D31449FCD81D1E734031D31449FCD81D&rvsmid=3EB5DE52055C
55C171153EB5DE52055C55C17115&FORM=VDRVRV

Spin Coating:

Spin coating is a procedure used to deposit uniform thin films to flat substrates. Usually a small amount of coating material is applied on the center of the substrate, which is either spinning at low speed or not spinning at all. The substrate is then rotated at high speed in order to spread the coating material by centrifugal force.

Spin coating is one of the most common techniques for applying thin films to substrates. It is used in a wide variety of industries and technology sectors. The advantage of **spin coating** is its ability to quickly and easily produce very uniform films, ranging from a few nanometers to a few microns in thickness.

Recommended videos:

https://www.bing.com/videos/search?q=Spin+Coating&&view=detail&mid=14A
E0450D07029C5FD0714AE0450D07029C5FD07&&FORM=VRDGAR

https://www.bing.com/videos/search?q=Spin+Coating&&view=detail&mid=D4E
F750DE5FECD316F01D4EF750DE5FECD316F01&&FORM=VDRVRV

Alignment and Exposure:

Optical lithography (photolithography) is a key process in microfabrication for transferring a geometrical pattern to a thin film substrate. The optical lithography mainly includes: wafer cleaning, photoresist **spin-coating**, soft backing, mask **alignment and**

exposure, photo resist developing, hard backing, etching and finally resist removal. A photomask is **aligned** and placed on the coated wafer with precision instruments. The wafer and **mask** are then **exposed** to ultraviolet (UV) radiation from an intense mercury arc lamp. This causes **exposure** to the photo resist in places not protected by opaque regions of the **mask**.

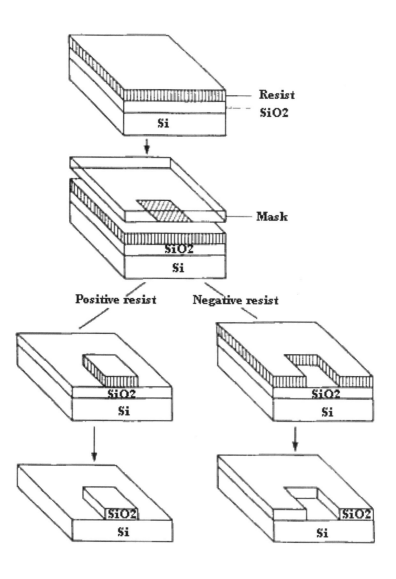

The contact aligner is a tool that performs **alignment** and **exposure** of wafers. The features on the contact aligner **mask** are the same size as they should be on the wafer (i.e. 1x magnification). Pattern transfer takes place by printing, i.e. by placing the **mask** in direct contact with the surface of the wafer, and then exposing it to UV light.

MASK ALIGNMENT & EXPOSURE

Transfers the mask image to the resist-coated wafer

Activates photo-sensitive components of photoresist

Three types of masking
(1) Contact printing
(2) Proximity printing
(3) Projection printing

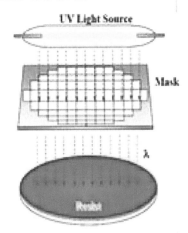

Recommended videos:

https://www.bing.com/videos/search?q=Alignment+and+Exposure&&view=detail&mid=51A0586EA48DBD5F8FFE51A0586EA48DBD5F8FFE&&FORM=VRDGAR

https://www.bing.com/videos/search?q=Alignment+and+Exposure&&view=detail&mid=DAA23E4E9AC2254F1A76DAA23E4E9AC2254F1A76&&FORM=VDRVRV

Development:

Development is the dissolution of unpolymerized resist. After exposure to UV light, the wafer is rinsed in a developing solution removing areas of photoresist and leaving behind a pattern of bare and photoresist coated regions on the wafer. Depending on the type of photoresist used, the photoresist will undergo one or two possible transformations when exposed to UV light. When light illuminates a positive photoresist, the exposure regions break down and become more soluble in a developing solution. As a result, the exposed photoresist can be removed when immersed in developing solution. A negative photoresist becomes cross-linked when exposed to UV light, becoming insoluble in the developing solution. Upon contact with developing solution, regions not exposed to UV light will be removed.

Lift-off is most commonly employed in patterning metal films for interconnections. After exposure to the developer solution, the wafer is coated with a thin layer of metal Afterwards, an appropriate solvent (such as acetone) is used to remove the remaining parts of the photoresist and the deposited film atop these parts of the resist can be lifted off.

Recommended videos:

https://www.bing.com/videos/search?q=Photolithography+Development&&view=detail&mid=41C6B2505098C7B9B3DB41C6B2505098C7B9B3DB&&FORM=VRDGAR

https://www.bing.com/videos/search?q=Photolithography+Development&&view=detail&mid=C47CCDDEFB7CA094F1C0C47CCDDEFB7CA094F1C0&&FORM=VRDGAR

Nanotechnology of the future.

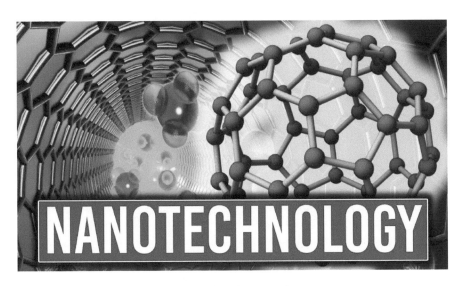

Nanotechnology today.

References:

1) "A Laboratory Course in Nanoscience and Nanotechnology", Dr. Gerrard Eddy Jai Poinern, *CRC Press, 2015;*

2) "Essentials in Nanoscience and Nanotechnology", Narendra Kumar and Sunita Kumbhat, John *Wiley & Son, 2016*;

3) "Analysis of lithography-based approaches in development of semiconductors". Chopra, J., 2014, *Int. J. Comput. Sci. Inf. Technol. Adv. Res. 6: 61-72*;

4) "The Potencial Environmental Impacts of Engineered Nanoparticles", *Nat. Biotechnol., 21 (10), (2003) 1166-1170*;

5) "Nanotechnology and Nanomaterials in Treatment of Life-Threatening Diseases", N. Kumar & R. Kumar, *Elsevier, 2014*;

6) "Basic Principles of Nanotechnology", Wesley C. Sanders, *CRC Press, 2019*;

7) "Micro- and nanofabrication methods in nanotechnological medical and pharmaceutical devices", T. Betancourt and L. Brannon-Peppas, *Int. J. Med. Sci. 1:483-495, 2006*;

8) "Fullerene Derivatives: An Attractive Tool for biological Applications"' S. Bosi, T. da Ros, Spalluto G and M. Prato, *Eur. J. Med. Chem. 38 (2003) 913-923*;

9) "Cleaner Nanotechnology and Hazard Reduction of Manufactured Nanoparticles", L. Reijnder, *J. Clean. Prod. 14 (2006) 124-133*;

10) "Fabrication of polyvinylpyrrolidone micro-/nanostructures using microcontact printing". Sanders, W.C., 2015, *J. Chem. Educ. 92: 1908-1912*;

11) "Health and Environmental Impact of Nanotechnology: Toxicological Assessment of Manufactured Nanoparticles", K. Dreher, Toxicol. Sci., 77 (1), (2004) 3-5;

12) Nanotechnology and Nanoparticles in Drug Delivery

https://www.understandingnano.com/nanotechnology-drug-delivery.html;

13) Nanotechnology-based drug delivery systems

https://www.ncbi.nlm.nih.gov/pmc/articles/PMC2222591/;

14) The University of New Mexico Learning Modules, **www.scme-nm.org;**

15) Wikipedia.com;

16) Britanica.com;

NOTES:

NOTES:

NOTES:

NOTES:

NOTES: